Wireless

Transformations: Studies in the History of Science and Technology
Jed Buchwald, general editor

Sungook Hong, *Wireless: From Marconi's Black-Box to the Audion*

Myles Jackson, *Spectrum of Belief: Joseph von Fraunhofer and the Craft of Precision Optics*

William R. Newman and Anthony Grafton, editors, *Secrets of Nature: Astrology and Alchemy in Early Modern Europe*

Alan J. Rocke, *Nationalizing Science: Adolphe Wurtz and the Battle for French Chemistry*

Wireless

From Marconi's Black-Box to the Audion

Sungook Hong

The MIT Press
Cambridge, Massachusetts
London, England

First MIT Press paperback edition, 2010
© 2001 Massachusetts Institute of Technology

Set in Sabon by The MIT Press.

Library of Congress Cataloging-in-Publication Data

Hong, Sungook.
Wireless : from Marconi's black-box to the audion / Sungook Hong.
p. cm. — (Transformations: studies in the history of science and technology)
Includes bibliographical references and index.
ISBN 978-0-262-08298-3 (hc. : alk. paper) 978-0-262-51419-4 (pb. : alk. paper)
1. Radio—History. I. Title. II. Transformations (MIT Press)
TK6547 .H66 2001
384.5'2'09034—dc21   2001030636

to my family

Contents

Preface

As late as 1850 there was no awareness of electromagnetic waves. In the 1860s, the British physicist James Clerk Maxwell theorized the existence of electromagnetic disturbances in the ether whose wavelengths were longer than those of infrared radiation. The followers of Maxwell who came to be known as the British Maxwellians concentrated on producing electromagnetic waves. In 1882–83, George FitzGerald and Oliver Lodge concluded that Maxwell's electromagnetic waves could be emitted from rapid electric oscillations produced by the discharge of a condenser or a Leyden jar, but they lacked an adequate device with which to detect this radiation.

In 1887–88, while Lodge was working on the oscillatory effect along wires, the German physicist Heinrich Hertz, working under the influence of Hermann von Helmholtz, succeeded in producing and stabilizing a new effect of sparks, which he soon identified as electromagnetic waves. Within a year, the British Maxwellian Oliver Heaviside exclaimed: "Not so long ago [electromagnetic waves] were nowhere; now they are everywhere!" (Heaviside 1892, volume II, p. 489) Scientists and engineers began speculating more concretely about the possibility of wireless communication, long dreamed of by many.

In 1892, William Crookes offered a futuristic vision of wireless communication in a popular article (Crookes 1892). In 1895–1897, Hertzian wave telegraphy practical enough for communication was invented and patented by Guglielmo Marconi. Marconi's achievement inspired others; in 1897, for example, the British electrical engineer William Ayrton imagined a future in which anyone could call a friend "in an electromagnetic voice" and the friend could reply "I am at the bottom of the coal mine, or crossing the Andes, or in the middle of the Pacific" (Ayrton 1897). Soon amateur wireless operators were communicating by means of wireless telegraphy, and by 1910 a young amateur wireless operator, Edwin Howard Armstrong, could note casually: "For the last two weeks I have been calling A. P. Morgan with my 2 K.W., but have not yet picked him up. I think he is probably out of town as I used to hear his 3 K.W. every night a few months ago." (E. H. Armstrong to Mr. Underhill, 2 Dec. 1910, Armstrong Papers, Columbia University) Fifteen years after Marconi's invention, wireless telegraphy had become an essential means of communication, as well as a hobby for many.

The popular history of wireless telegraphy is a history of adventure. Many of us, as children, read with wonder the story of an Italian teenager's experiment that went unrecognized by his father, his move to Britain at 22, the accidental breakage of his instrument by British customs, the "secret-box" that astonished famous British scientists, his reception of transatlantic messages with a kite at age 27, and the heroic role of wireless telegraphy in the *Titanic* disaster. This story's sequel is the history of radio, in which David Sarnoff—an uneducated immigrant boy who admired Marconi and who started his career doing menial work at the American Marconi Company—eventually became the emperor of broadcasting at RCA.

Many of us (at least, many of us born before the advent of personal computers) had the experience of fabricating a radio from a kit consisting of a coil, wires, a tiny crystal, and an earphone. Some may remember the feeling of wonder when radio signals were first captured by such a primitive receiver. Similar feelings were experienced by the wireless pioneers. For example, in 1934, John Ambrose Fleming, the inventor of the thermionic valve (i.e., the diode vacuum tube), still remembered how surprised he had been in 1898 when he had seen a Morse printer printing a message from Marconi (Fleming 1934, p. 116).

The early development of wireless communications has been described in detail by Hugh Aitken, but even he did not attempt to probe the substance and context of scientific and engineering practice in the early years of wireless. In this book I attempt to fill that gap. My main goal is to provide a detailed analysis of engineering and scientific practices that are not only experimental but also theoretical. Examples include the engineering practices involved in Marconi's earliest wireless telegraphy, Fleming's "grafting" of power technology onto wireless telegraphy, Marconi's innovation in tuning as represented in his "four-seven" patent, Fleming's research on the unilateral conductivity in the Edison effect, and Lee de Forest's invention of the triode. I also aim to explore the borderland between science and technology. For that purpose I pay close attention to the work of Fleming, who was trained in experimental physics by Maxwell but who moved to wireless telegraphy when he became Marconi's scientific advisor. One important mode of such "mediation" is the transformation of scientific effects into technological artifacts. I discuss several notable cases of this process, including the transformation of Hertzian waves into practical wireless telegraphy, Marconi's transformation of a resonance principle into a stable artifact, Fleming's utilization of the unilateral conductivity for the valve, and the transformation of "negative resistance" into the transmitter of continuous waves. Taken together, my accounts of these engineering practices provide a window on or a context for such novel practices. (By context I mean techno-scientific, corporate, or authorial settings that promoted or constrained inventions and innovations.) My analyses of the Maskelyne affair, of tuning technology, and of the invention of the valve in the context of corporate politics exemplify my efforts to link engineering practices with specific contexts. The construction of these contexts is based largely on archival sources previously overlooked or underused by historians.

In chapter 1, I discuss the origins of wireless telegraphy by contrasting physicists' Hertzian optics with Marconi's telegraphic imperatives. I classify the scientists and engineers who claimed to have devised wireless telegraphy before Marconi into three categories: (1) British Maxwellian physicists and engineers, including John Perry, George FitzGerald, Frederick Trouton, Alexander Trotter, and Richard Threlfall, who suggested a practical use of Hertzian waves for communication in the early 1890s, (2) William Crookes, who provided a detailed description of wireless telegraphy in 1892, and (3) Ernest Rutherford and Captain Henry Jackson, who actually performed experiments for communication with Hertzian waves. I show that all this work was constrained by optical analogies of electromagnetic waves. These researchers had access to a complex web of conceptual and instrumental resources associated with Hertzian optics; however, these optical resources, which opened a new arena of physical research and even opened new technological possibilities, also conditioned and constrained the transformation of Hertzian apparatuses into practical wireless telegraphy. Re-evaluating Marconi's ingenuity in a more historical and contextual manner, I discuss how Marconi's ingenuity and originality sprang from his devotion to perfecting the homology between wired and wireless telegraphy.

In chapter 2, I take up the priority dispute between Marconi and Oliver Lodge, who (it has been claimed) demonstrated wireless transmission of alphabetic messages in 1894. I show that Lodge's putative demonstration of wireless telegraphy in 1894 had nothing to do with telegraphy, alphabetic signals, or dots and dashes. Why, then, did Lodge and others claim priority? In considering this question, I discuss the impact of Marconi and his British patent on the Maxwellian physicists, particularly Lodge, Silvanus Thompson, FitzGerald, and Fleming. Marconi was known to them as an Italian "practician," yet he was the one who achieved the transformation of Hertzian laboratory apparatus into commercial wireless telegraphy. Marconi's patent was so strong and overarching that it seemed to monopolize Hertzian waves and thereby encroach upon British national interest. The image of Lodge as the inventor of wireless telegraphy was deliberately constructed by Lodge's friends and by Lodge himself to counter the effects of Marconi's priority and his strong patent.

Between 1897 and 1901, Marconi gradually increased the transmitting range of his apparatus to 200 miles, which was adequate for commercial ship-to-shore communication. In December 1901, he succeeded in transmitting across the Atlantic. Chapter 3 highlights the role of Fleming, Marconi's scientific advisor, in achieving this rapid transformation of telegraphic devices into a power system. This chapter is based on a detailed analysis of Fleming's unpublished notebooks and other manuscript sources. Using these archival sources, I analyze Fleming's laboratory experiment to "graft" power engineering onto wireless telegraphy between July and December 1900. I then turn to Fleming's experiments in the field between January and September 1901, which were essential to the project's success. I show that Fleming used these experiments to raise his status and credibility in the Marconi Company. However, Marconi's intervention in the project in the summer of 1901 was also crucial, and it displaced Fleming from the experiments. Through detailed examination of Fleming's and Marconi's work, I compare their different "styles" of engineering, and I discuss how these different styles clashed in their transatlantic experiment.

In chapter 4, I examine the "Maskelyne affair" of 1903, in which Nevil Maskelyne interfered with Fleming's public demonstration of Marconi's syntonic (tuning) system at the Royal Institution by sending derogatory messages from his own transmitter. This affair severely damaged Marconi's and Fleming's credibility, insofar as witnessing and reporting successful experiments constituted an important strategic resource for them. Indeed, soon after the Maskelyne affair, Fleming, who had been a trustworthy witness to Marconi's secret demonstration of the efficacy of his system, was dismissed from his advisory post. A detailed analysis of the affair uncovers several issues that are worth exploring, including the struggle between Marconi and his opponents, the efficacy of early syntonic devices, Fleming's role as a public witness to Marconi's private experiments, and the nature of Marconi's "shows." In addition, the Maskelyne affair provides a rare case study of how the credibility of engineers was created, consumed, and suddenly destroyed.

Fleming regained his scientific advisorship to Marconi in 1905 after inventing the thermionic valve, which rectified the high-frequency alternating currents induced in a receiving antenna by electromagnetic oscillations and could therefore be used as a detector. Chapter 5 revises the well-known story of the valve, partly because the valve was the "stem cell" for the entire body of various vacuum tubes but also because its invention was dramatic. Thomas Edison discovered a curious effect (which came to be called the Edison effect) in the early 1880s, but he did not understand its mechanism. Fleming, a physicist-engineer working for the British Edison Company, investigated the effect in the 1880s and the 1890s. In 1904, while serving as scientific advisor to Marconi, he used his theoretical understanding of the effect to create the thermionic valve, which he intended as a signal detector for wireless telegraphy. Although this popular story was first told by Fleming and has been widely known since then, I aim to provide an alternative narrative that corrects three points. First, I re-examine the relationship between the Edison effect and the thermionic valve in the Maxwellian context in which Fleming's research on the Edison effect was originally performed. I show that Fleming's examination of the Edison effect was essential to the invention of the valve, not because of the essential character of the effect itself, but because of the uniquely Maxwellian context in which Fleming conducted his research. That context led him to devise a circuit in which an Edison lamp, a probe, a battery, and a galvanometer were uniquely combined, and it was this circuit that Fleming transformed into the thermionic valve in 1904. Second, I show that the termination of his scientific advisorship to the Marconi Company in December 1903 impelled Fleming to change his research style, and that his efforts to regain credibility at the Marconi Company were crucial to his invention of the valve.

Third, I argue that Fleming actually intended the valve to be a high-frequency alternating current measurer for use in the laboratory, and that Marconi, not Fleming, transformed it into a practical receiver.

In chapter 6, I focus on the audion and on the shift from wireless telegraphy to radio. Fleming's valve was an inspiration to the American scientist and inventor Lee de Forest, who invented the grid audion in 1906. In the 1910s, it was discovered that the audion could be used as a feedback amplifier and as an oscillator. As an amplifier, the audion was several hundred times more sensitive than ordinary detectors. As an oscillator, it made producing continuous waves—and transmitting the human voice—simple and cheap. I first describe how de Forest invented the grid audion at the end of 1906. In particular, I discuss the influence of Fleming's valve on the audion. I also offer a new view of the history of continuous waves—a view that traces a reasonably continuous path from the "negative resistance" of the electric arc through the arc generator to the audion. To accomplish this, I explore a theoretical and instrumental continuity from the arc's negative resistance to the oscillating audion, a continuity that was most apparent in the fact that each hissed or whistled.

Acknowledgments

I would like to express my sincere thanks to Jed Buchwald, who always inspired me with novel ideas and insights. My colleagues and friends who read earlier versions of chapters 2–5 (Hong 1994c, 1996a–c) made valuable comments and suggestions that are reflected in the book. Special thanks to Bill Aspray, Bernard Carlson, Hasok Chang, Jonathan Coopersmith, Berhard Finn, Peter Galison, Charles Gillispie, Yves Gingras, Graeme Gooday, Bert Hall, Takehiko Hashimoto, Bruce Hunt, Ed Jurkowitz, Yung Sik Kim, Janis Langins, Trevor Levere, Donald MacKenzie, Rik Nebeker, Henry Paynter, Robert Post, Roy Rodwell, Hugh Slotten, Thad Trenn, and Ido Yavetz. I am also grateful to the late Gioia Marconi Braga for permitting me to quote Guglielmo Marconi's private letters from her collection. I thank the Library of University College London for permitting me to quote from documents in the archives of the Fleming Collection and the Lodge Collection. I am also thankful to Marconi plc and the IEE Archives for their generous help. My research has been generously supported by the Canadian Social Sciences and Humanities Research Council through research grant 410961236, for which I am grateful.

Wireless

1

Hertzian Optics and Wireless Telegraphy

If I could get an appreciable effect at ten miles, I would probably be able to make a considerable amount of money out of it.
—Ernest Rutherford, letter to his family, January 1896 (Eve 1939, p. 22)

Why Do the Origins of Radio Matter?

In 1995 the centennial of the invention of radio by Guglielmo Marconi (1874–1937) was celebrated around the world. Exhibitions, symposia, and itinerant lectures immortalized the inventor and his invention. Not everyone, however, was content. Several British historians and engineers had commemorated 1994 as the centennial of the invention of wireless telegraphy—not by Marconi, but by the noted British physicist Oliver Lodge (1851–1940).[1] Attempts by the Soviet Union in the 1950s and the 1960s to get Aleksander Popov recognized as the true inventor of wireless telegraphy are well known. A careful study of Popov by the historian Charles Süsskind revealed that the Soviets' claim was based not so much on historical evidence as on politics. When claims for Popov resurfaced in 1995, they were rebutted—but on the behalf of Lodge, not Marconi.[2]

The invention of wireless telegraphy has been a subject of controversy for more than 100 years. In 1897, Lodge wrote in *The Electrician* of British scientists and engineers who had worked on wireless telegraphy before Marconi. In 1896, Marconi had arrived in England, had contacted the Chief Electrician of the Post Office, William H. Preece, and had filed his first patent on Hertzian-wave telegraphy. If Marconi's was the first such patent, how could Lodge have recalled any work on wireless telegraphy before Marconi? Lodge, however, referred to two kinds of antecedents. First, before Marconi, several individuals had actually devised something that could, in retrospect, be called wireless telegraphy. These people did not widely announce their inventions, nor did they apply for a patent for them, perhaps because they hardly considered their inventions to be practical. Second, the principles of wireless telegraphy—sending and receiving Morse-coded messages through Hertzian waves—were already known and had already been put into practice. On these bases, Lodge claimed that George M. Minchin, Ernest Rutherford (then a student of J. J. Thomson), and Lodge himself had worked on wireless telegraphy before Marconi. Lodge also added the name of William Crookes, who "indeed, had already clearly stated this telegraphic application of Hertz waves in the *Fortnightly Magazine* for February 1892" (Lodge 1897, p. 90). In 1900, Lodge added to the list Alexander Muirhead, Henry Jackson, J. Chunder Bose, Augusto Righi, and Aleksander Popov (Lodge 1900, pp. 45–49).

In this chapter, I will critically analyze Lodge's claim that several individuals devised wireless telegraphy before Marconi. I will examine the cases for British Maxwellian physicists and engineers, including John Perry, George FitzGerald, Frederick Trouton, Alexander Trotter, Richard Threlfall, William Crookes, Ernest Rutherford, and Henry Jackson. I will show that their research and their conceptions were largely constrained by their optical analogies for electromagnetic waves. I will also reveal how, before Marconi,[3] a complex web of conceptual and instrumental resources opened a new arena of physical research and technological possibility, yet at the same time conditioned and constrained possibilities for the transformation of Hertzian apparatuses into practical wireless telegraphy. I will then use this complexity in re-evaluating Marconi's ingenuity. A detailed explanation of how Marconi proceeded step by step to transform Hertzian apparatuses into practical wireless telegraphy will be provided. I will show that Marconi's ingenuity and originality sprang from his devotion to perfecting the homology between wired and wireless telegraphy.

Experimental Expansions and Theoretical Constraints, 1888–1896

Heinrich Hertz produced controllable electromagnetic waves in the laboratory late in 1887 (figure 1.1). Many physicists rapidly pursued the study of the wavelike effects associated with it. In the early 1890s, determining such optical and electromagnetic properties of a material as its specific inductive capacity, coefficient of absorption, and index of refraction by

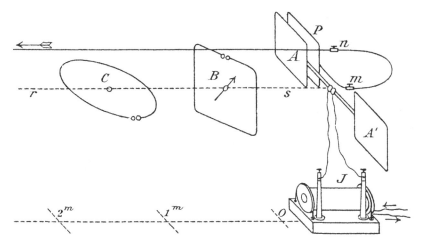

Figure 1.1
Heinrich Hertz's linear spark-gap oscillator and ring-shaped spark gap resonator.
A and A' are condenser plates; J is an induction coil.

means of the Hertzian wave and probing the optical characteristics of the wave itself constituted a major research program. In addition, some physicists, including some of the British Maxwellians, used Hertzian waves for *instrumental* purposes. For example, J. J. Thomson, then Cavendish Professor of Experimental Physics at Cambridge, used the wave to create glow discharge in vacuum tubes in order to examine the nature of conduction in a near vacuum (Thomson 1891), Oliver Lodge (1890a) compared Hertz's electric resonator's reception of waves to the human eye's reception of light, and the physicist George Minchin (1891) used the waves to stimulate his solar cell.

Hertz's device was essentially a combination of a transmitter and a receiver. The linear, dumbbell-shaped transmitter consisted of chemical batteries, an induction coil, an interrupter, and two condenser plates connected by a linear wire with a spark gap in the middle. Electrostatic energy was stored in the condenser plates by the high voltage created through the action of the induction coil. This energy, which could not be stored in the plates indefinitely, was released into space in the form of electromagnetic waves when a spark was formed across the gap. Simply put, an electromagnetic wave is always created when one can see a spark (even on a cat's hair). Hertz's receiver, which he called a resonator, was a circular wire with a tiny spark gap. When an electromagnetic wave fell on it, its energy was transformed into a high-frequency alternating current that caused a spark in the gap (Buchwald 1994).

After Hertz discovered electromagnetic waves, new forms of transmitters and receivers were invented. Spherical transmitters were used widely. Innovations in the receiver were more remarkable. In particular, the "coherer" almost universally replaced Hertz's resonator as a receiver. The history of the coherer is of some interest here. In 1890, while experimenting on lightning rods, Lodge found that two metallic conductors separated by a tiny air gap were fused when an oscillatory discharge passed through them. At that time, Lodge accepted David Hughes's explanation that this was a thermoelectric phenomenon, and dropped the subject. In 1890, in France, Edouard Branly found that fine copper filings in a glass tube conducted feebly under ordinary conditions, but that their conductivity increased abruptly when a spark was generated nearby. In 1891 the British journal *The Electrician* published full translations of Branly's articles, complete with figures, but they were apparently overlooked. They were noticed later, however, by the British physicist Dawson Turner, who demonstrated the decrease in the resistance of copper filings at a meeting of the British Association for the Advancement of Science in Edinburgh in 1892. Turner's demonstration was seen by the physicist W. B. Croft, who conducted a short experiment on the same phenomenon before the Physical Society of London in October 1893. There, George Minchin of the Royal School of Engineering noticed the similarity between Croft's (actually Branly's) tube and his solar cell's response to Hertzian waves. Minchin immediately read a paper on the subject at a meeting of the Physical Society. While hearing Minchin's paper, Lodge noticed that Branly's and Minchin's discovery was very similar to his previous research on the action of lightning in a tiny metallic gap. Lodge reasoned that electromagnetic radiation made the metallic molecules in the filings and those in the microscopic air gap cohere with one another. Based on this similarity, Lodge soon devised a single-point-contact coherer with a spring wire and an aluminum plate. Lodge soon found that the coherer was not only much more sensitive than ordinary spark-gap resonators but also more sensitive than Branly's filing tube (figure 1.2).[4] Such Hertzian devices became standard equipment in physics laboratories.

Maxwellian physicists did not, however, remain satisfied with the instrumental use of Hertzian waves. The idea of using the wave for communi-

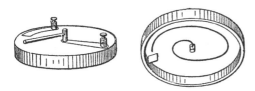

Figure 1.2
Above: The Branly tube as improved by Lodge. Below: Lodge's single-contact coherer. Source: Lodge 1894a, p. 337.

cation purposes also captured their imagination. For instance, John Perry, a Maxwellian physicist and electrical engineer, predicted the practical use of Hertzian waves in 1890 as follows (Perry 1910, p. 117):

> . . . we now know, from the work of Professor Hertz, that . . . this now recognized kind of radiation may be reflected and refracted, and yet will pass through brick and stone walls and foggy atmospheres where light cannot pass, and that possibly all military and marine and lighthouse signalling may be conducted in the future through the agency of this new and wonderful kind of radiation, of which what we call light is merely one form. Why at this moment, for all I know, two citizens of Leeds may be signalling to each other in this way through half of a mile of houses, including this hall in which we are present.

Yet it was not easy to make practical use of Hertzian waves. First of all, the obvious limit in the wave's transmitting distance was a solid barrier to any kind of progress. In 1891, Minchin detected the wave at a distance of 130 feet from the source by using his photocell as a detector. He had great difficulty detecting it 150 feet away. In 1894, Lodge detected the Hertzian wave at a distance of 70 yards, and even that was the maximum distance with his most sensitive detector. Several factors contributed to such short transmitting distances. Above all, neither Minchin's nor Lodge's main interest lay in increasing the transmitting distance. Minchin was interested in testing the response of his solar cell to Hertzian waves. Once he discovered the cell's response to weak electromagnetic waves, he no longer pursued this topic. Lodge was mainly interested in examining the mechanism of optical recognition of human eyes by an analogy with the coherer. Lodge's "long-distance" experiment was performed to simulate the detection of very weak light by human eyes.[5]

The social space in which both Minchin and Lodge performed their experiments helped consolidate and naturalize the limit. From the beginning, Hertzian devices—such as the induction coil, the spark transmitter, and the resonator—were laboratory apparatuses. Seldom were all these devices moved out of the laboratory together. For "long-distance" trials, Lodge moved his transmitter to the Zoology Theatre at the University of Liverpool and left the coherer at his Physics laboratory. He moved the transmitter because it was less affected by movement than the sensitive detector. Minchin instead moved his detector—consisting of a solar cell, a galvanometer, and a wire (which he called the "feeler") that connected them—from his laboratory through a lecture room to another laboratory. One day he took the detector out of the building and across a lawn to the tennis courts at the Royal School of Engineering, where he found that the wire netting of the court partly blocked the wave. "Unfortunately," the physicist John Trowbridge wrote in 1897, "at present we cannot detect the electromagnetic waves more than 100 feet from their source" (Trowbridge 1897, p. 256).

Theoretical inferences exacerbated the situation. FitzGerald's (1883) theoretical formula for determining the energy of radiation emitted from a closed circuit indicated that, other conditions being equal, the energy of radiation was inversely proportional to the fourth power of the wavelength. That is, the shorter the wavelength, the greater the energy of the radiation, and thus the higher the possibility of being detected at a distance. Although FitzGerald's formula was devised for a closed circuit, distinctions between closed and open circuits were often ignored. In 1894, FitzGerald's close friend Lodge, who had already moved in the direction of decreasing the wavelength for optical purposes, justified his use of such a short wave for a "long-distance" trial on the grounds of FitzGerald's formula and the measured sensitivity of the best detector. Lodge estimated the wave's maximum distance of transmission to be half a mile, although he appended "this is a rash statement not at present verified."[6]

FitzGerald, a professor of natural and experimental philosophy at Trinity College Dublin, was troubled by another theoretical predicament: the damping of Hertzian waves. In December 1891, Frederick Trouton, FitzGerald's number-one assistant, stated in a public lecture that "no doubt [electric waves] will be turned by man to subserve useful purposes," but that their insufficient power caused by the high damping of spark-generated waves was a barrier to overcome. The only way to avoid damping was to create a powerful *continuous* wave. Trouton (1892, p. 303) wrote:

The great difficulty which lies in the way is, that the radiation produced by the automatic method with intermittent supply of energy, the only one yet used, is necessarily insufficient in intensity to be powerful at considerable distances, owing to the vibration dying rapidly out, like that of a violin string. So that until some other form of apparatus is invented, to be supplied continuously with energy, so as to produce a continuous vibration, and to which we can harness on a 50 H.P. engine, and get a really powerful beam of these electric rays, there can be no great advance hoped for. . . . There is little doubt but that a powerful beam of this sort would, unlike light, be unabsorbed by fog; so looking into the future, one sees along our coasts the light-house giving way to the electric house, where electric rays are generated and sent out, to be received by suitable apparatus on the passing ships, with the incomparable advantage that at the most critical time—*in foggy weather*—the ship would continue to receive the guiding rays.

Trouton's emphasis on diminishing damping was based on FitzGerald's previous research on damping. FitzGerald had recognized that radiation energy created from spark discharge was sent off in a portion of the interval between consecutive sparks ranging from a hundredth to a thousandth of the interval. Because of this, the energy is shot off almost like a pulse rather than in a continuous form. As a result, the resonator is not stimulated often enough to keep from dying away at a great distance, where the pulse is naturally weak anyway. What was needed here was something like an "electric whistle" that could produce a strong beam of continuous waves, just as an ordinary whistle produces continuous sound. In 1892, for this purpose, FitzGerald (probably working with Trouton) tried to construct a continuous-wave transmitter. FitzGerald's transmitter utilized the "adjuvant electromotive force" or "negative resistance" of a working dynamo, but it failed to generate a continuous wave that could store more energy than the damped wave (FitzGerald 1892).

Most important, the Maxwellian physicists' adherence to optics obscured a telegraphic application of Hertzian waves. Even when they imagined signaling by means of Hertzian waves, an optical rather than a telegraphic analogy dominated. It was partly because the similarity of the Hertzian apparatus to light signaling devices (such as a lighthouse or a heliograph) was more conspicuous than any similarity to telegraphic technologies. Light signaling devices consisted of a light source (lamps) and a detector (human eyes), which corresponded to a Hertzian transmitter and a detector (the resonator or coherer). There were good theoretical reasons for this analogy. According to James Clerk Maxwell, the only difference between light and Hertzian waves was their wavelengths; according to William Thomson (Lord Kelvin), Hertz's resonator was an "electric eye"; according to Lodge, human eyes were quite similar to the coherer.[7] In contrast, there was little in common between telegraphic devices (a Morse receiver and a long wire) and the Hertzian apparatus. Still worse, there was an important asymmetry between them, in that there was no earth-return (i.e., the assumed flow of current through the earth) in the latter. An earth-return was essential in cable telegraphy.

This conformity to light signaling is found in the work of several physicists and engineers with a Maxwellian inclination, including Richard Threlfall and A. P. Trotter. At the annual meeting of the Australasian Association for the Advancement of Science in 1890, Threlfall, who was then professor of physics at the University of Sydney and who had been J. J. Thomson's student at the Cavendish Laboratory, also spoke of "a sort of ray flasher" for Hertzian waves. While discussing Maxwell's electromagnetic theory of light and Hertzian experiments, Threlfall (1890, p. 46) stated: "If it be permissible to prophesy wildly, we may see in this observation [the detection of Hertzian waves by means of a Geissler tube] the germ of a great future development. Signaling, for instance, might be accomplished secretly by means of a sort of electric ray flasher, the signals being invisible to anyone not provided with a properly turned tube." The optical analogy was most evident in the remarks of Trotter, the editor of *The Electrician* in the 1890s. In March 1891, when a shipwreck caused by thick fog that screened the light of St. Catherine's lighthouse created a public sensation, Trotter was inspired to suggest, in the editorial column of *The Electrician*, that Hertzian waves be used for ship-to-shore communication on foggy days. According to Trotter, if waves with quite a short wavelength (say, 1 millimeter) could be produced, their enormous energy would make "these radiations . . . pierce not only a fog, but a brick wall." Trotter proposed Hertzian-wave "flash signals" between lightships and the shore.[8] Signaling with Hertzian waves would be similar to light signaling, in which a light source is screened at intervals. A telegraphic key was not likely to have been on his mind; even after Marconi's invention was publicized, an editorialist at *The Electrician* remarked: "No doubt, short and long electric impulses could be sent by mechanically interposing suitable electro-magnetic screens between the source of energy and the receiver." (*The Electrician* 37, September 25, 1896, p. 685).

Neither Threlfall's lecture nor Trotter's editorial was influential. Their prophesies were pioneering but purely speculative. Trotter's editorial note was widely advertised by *The Electrician* only after Marconi announced in 1897 that "until the date of my experiments no mention was made in the scientific papers of the possibility of long-distance signals being transmitted by Hertzian waves."[9] To my knowledge, Trouton's lecture and FitzGerald's trial were given no attention by their contemporaries. More important, none of their contemporaries seriously considered a telegraphic application. They were preoccupied with optics, not communication technology—and, as a consequence, with the light flasher, not telegraphy.

William Crookes's Wireless Telegraphy: Technology as Dream and Reality

Some scientists and engineers—not of the Maxwellian variety—did suggest telegraphic application of Hertzian waves. One of them was William Crookes, who held an unusual position in the late Victorian scientific community.[10] In the 1860s, he devised a vacuum pump that could produce a high vacuum in a glass tube. Using what we would call a cathode ray tube, he performed a series of well-known experiments on the "fourth state of matter," or "radiant matter." On the basis of these experiments, he invented the radiometer. Yet his theoretical concept, the fourth state of matter, was generally thought speculative.

In the February 1892 issue of the *Fortnightly Review*, Crookes published a paper that is now famous for having predicted the advent of wireless telegraphy. Not only did he mention wireless "telegraphy" rather than a flashing device; he also predicted radiotelegraphic features such as tuning and addressed the need for workable generators, for sensitive receivers with tuning, and for a means of directing waves (Crookes 1892, pp. 174–175):

Rays of light will not pierce through a wall, nor, as we know only too well, through a London fog. But the electrical vibrations of a yard or more in wave-length of which I have spoken will easily pierce such mediums, which to them will be transparent. Here, then, is revealed the bewildering possibility of telegraphy without wires, posts, cables, or any of our present costly appliances. . . . Also an experimentalist at a distance can receive some, if not all, of these rays on a properly-constituted instrument, and by concerted signals messages in the Morse code can thus pass from one operator to another. What therefore remains to be discovered is—firstly, simpler and more certain means of generating electrical rays of any desired wave-length, from the shortest, say of a few feet in length, which will easily pass through buildings and fogs, to those long waves whose lengths are measured by tens, hundreds, and thousands of miles; secondly, more delicate receivers which respond to wave-lengths between certain defined limits and be silent to all others; thirdly, means of darting the ahead of rays in any desired direction, whether by lenses or reflectors. . . .

He then added an interesting comment on wireless communication between two friends (ibid., p. 175):

Any two friends living within the radius of sensibility of their receiving instruments, having first decided on their special wavelength and attuned their respective instruments to mutual receptivity, could thus communicate as long as often as they pleased by timing the impulses to produce long and short intervals on the ordinary Morse code.

The aforementioned article has been appraised as "a remarkable prediction and a correct analysis of the principal obstacle that would have to be overcome before radiotelegraphy could become a reality" (Süsskind 1969a, p. 70). Furthermore, several scholars have emphasized the influence of Crookes's article on other pioneers of radio telegraphy. The historian Hugh Aitken commented (1976, p. 114):

Crookes's article was read very widely—and more than that, attended to and remembered—both in Europe and in the United States; there is hardly one figure important in the early days of radio who does not at some point in his memoirs or correspondence refer to the article of 1892 as having made a difference. . . . Crookes' article was both timely and catalytic. The year 1892 does mark a watershed. Before that, experimentation with electromagnetic waves was essentially a matter of validating Maxwellian theory; after, it became a matter of devising signalling systems, of inventions and patents, of developing a commercial technology.

Here Crookes is described as having paved the way for Lodge's wireless telegraphy in 1894, and for Popov's and Marconi's signaling in 1895. Even Marconi's former employee Gerald Isted made a similar comment (Isted 1991a, p. 50): "Without doubt, his paper was read and noted by . . . Branly, Preece, Lodge, Popov, Jackson and Marconi; from that point a great race to produce a workable system of 'wireless telegraphy' then commenced."

How much of Crookes's article was correct? From beginning to end, it was full of fantasies and futuristic stories. After mentioning wireless telegraphy, Crookes talked about using electricity to improve harvests, to kill parasites, to purify sewage, to eliminate disease, and to control weather. And there is a strange anachronism in his statement on wireless telegraphy. To the remark quoted above, he added: "This is no mere dream of a visionary philosopher"; however, he went on to say "even now, indeed, telegraphing without wires is possible within a restricted radius of a few hundred yards, and some years ago I assisted at experiments where messages were transmitted from one part of a house to another without an intervening wire by almost the identical means here described" (Crookes 1892, p. 176). Here Crookes appears to have regarded wireless transmission of messages as an accomplished fact in 1892. He later recalled that the experiment in which he said he assisted was David Hughes's experiment in 1879 on the reception of sparks, which Hughes called "aërial waves." Yet Hughes's experiment had in fact nothing to do with *signal* transmission.[11] Another possibility is that Crookes witnessed experiments on induction telegraphy (which is, of course, wireless), such as those conducted by Willoughby Smith in 1887, and that he confused electromagnetic induction with Hertzian oscillation.[12] In any case, it is most likely that he regarded his version of wireless telegraphy as a "sober fact."

Crookes's *Fortnightly Review* article was in fact an extension of his presidential speech at the meeting of the Institution of Electrical Engineers on November 13, 1891. This speech, published in *Nature*, was criticized severely in the magazine *The Spectator*: "It is really difficult to know whether we should understand [Crookes] literally, and take all his statements as the latest scientific truths." *The Spectator* even identified some of Crookes's stories as "fairy-tales of science."[13] *The Electrician* was not so critical as *The Spectator*, but it pointed out a few technical mistakes in Crookes's description of "ether vibrations."[14]

How influential was Crookes's article? My reading of various contemporary sources suggests that it was hardly noticed until 1897. One of the first instances in which its influence can be detected is an early 1897 lecture before the Imperial Institute in which William Ayrton repeated the "Crookesian" discourse of wireless communication between two friends (Ayrton 1897, p. 548):

There is no doubt the day will come, maybe when you and I are forgotten, when copper wires, gutta percha covering and iron sheathings will be relegated to the museum of antiquities. Then when a person wants to telegraph a friend, he knows not where, he will call in an electromagnetic voice, which will be heard loud by him who has the electromagnetic ear, but he will be silent to everyone else, he will call, "Where are you?" and the reply will come loud to the man with the electromagnetic ear, "I am at the bottom of the coal mine, or crossing the Andes, or in the middle of the Pacific. Or perhaps no voice will come at all, and he may then expect the friend is dead."

Yet Ayrton seems also to have been immensely inspired by practical successes of Marconi, which had been widely advertised in the last quarter of 1896. As a matter of fact, Crookes was hardly cited before the sensation that Marconi created. Lodge (1894b) did not mention Crookes in his Royal Institution lecture on the "Work of Hertz." In a series of Christmas Lectures at the Royal Institution in December 1896, Silvanus P. Thompson (1851–1916), who reviewed the history of Hertzian waves from Faraday and Maxwell to Marconi, did not mention Crookes either (S. P. Thompson 1897). In a memorandum submitted to the Marconi Company in 1904, Marconi's scientific advisor John Ambrose Fleming noted that Crookes's article "attract[ed] no notice for nearly five years until brought forward for the purpose of litigation."[15] *The Electrician* reprinted part of Crookes's 1892 article in its issue dated October 1, 1897,[16] just after Marconi's patent had been publicized. Its motive probably was to weaken Marconi's priority claim. It is not unlikely that Lodge came across Crookes's article only by reading the October 1, 1897 issue of *The Electrician*.

In any case, Oliver Lodge addressed the importance of Crookes in his article "History of the Coherer Principle," published in *The Electrician* of November 1897:

Numbers of people have worked at the detection of Hertz waves with filing tube receivers, and every one of them must have known that the transmission of telegraphic messages in this way over moderate distance was but a matter of demand and supply; Sir W. Crookes, indeed, had already clearly stated this telegraphic application of Hertz waves in the *Fortnightly Magazine* for February, 1892, and refers to certain experiments already conducted in that direction.

Here Lodge's intention seems to have been to propose that Marconi's invention was not a product of ingenuity but a product of the period. Before Marconi, Crookes was sneered at or neglected; after Marconi, Crookes was considered a visionary. If there is one thing common to Trotter, Threlfall, and Crookes, it is that they were all reinstated after 1897.

Why 1897? Marconi came to England in early 1896 and applied for his first patent (provisional specification) in the same year. In July 1897 the complete specification of Marconi's first patent on wireless telegraphy was accepted, and soon it began to be publicized. Around the same time, Marconi, seeking to establish his own company to exploit his patent, severed his connections to the British Post Office and to its Chief Electrician, William Preece. In his first patent for wireless telegraphy, Marconi claimed almost everything about the use of the coherer (which had been invented by Branly and improved by FitzGerald and Lodge) in wireless telegraphy. In May 1897, Lodge had applied a patent for a system of wireless telegraphy of his own (consisting of the closed transmitter, the tapper, the coherer receiver, and tuning between the transmitter and the receiver), but he had had to withdraw his claims on the coherer and the tapper because they had been so thoroughly covered by Marconi. The military and naval usefulness of wireless telegraphy gradually became obvious. If Marconi's patent were to go unchallenged, it would monopolize not only Hertzian waves but also important British national interests. Lodge argued that George Minchin, William Crookes, Ernest Rutherford, and Lodge himself preceded Marconi. J. J. Thomson, Minchin, Rollo Appleyard, and Campbell Swinton joined Lodge in undermining Marconi's claim to originality.[17]

From Effect to Artifact: Rutherford and Jackson

Although Marconi's 1896 patent was the first patent on Hertzian-wave telegraphy, Marconi was not the only person who utilized Hertzian waves for telegraphy. In 1895–96, quite independent of Marconi, Ernest Rutherford and Captain Henry Jackson were doing some experiments with the application of Hertzian waves to telegraphic communication in mind. The similarities and differences between them, as well as those between them and Marconi, are illuminating.

In early October 1895, Rutherford, who had just come from New Zealand to the Cavendish Laboratory, began his research on the effect of electromagnetic waves on a piece of magnetized iron—a topic on which he had worked in New Zealand. Having transformed a magnetized needle into a numerical detector of Hertzian waves by connecting a galvanometer to it, he performed several experiments to measure quantities such as the damping of a wave, the capacitance of a capacitor, and the wavelength of a wave on the wire. For these experiments, Rutherford used a closed-type transmitter; it was a good vibrator but a bad radiator. At the end of 1895, he tried other vibrators. With a spherical radiator, he found it difficult to obtain a large effect. With "the ordinary form of the Hertzian dumb-bell vibrator," he observed a large deflection in his detector at a distance of 50 feet. Rutherford then moved the vibrator to the lecture room, 40 yards away, and still observed the deflection. The wave apparently "pass[ed] through 3 or 4 walls and two floors in its passage."[18] This kindled Rutherford's curiosity, since the sensitivity of his new detector was one of his prime concerns. He constructed huge plates (6 feet by 3 feet) for the vibrator, set them up on the second floor of the Cavendish Laboratory, and moved the detector to James Ewing's Engineering Laboratory, more than 90 yards away in an adjacent building. Between the two labs were "two thick walls and pipes running along the walls" (Rutherford, *Laboratory Notebook*, pp. 78–79). (See figure 1.3.) Despite this, a large deflection was obtained. J. J. Thomson, the director of the Cavendish Laboratory, became interested in Rutherford's research and pushed him to find how far he could transmit. Rutherford changed the settings of his magnetic-needle detector to obtain the most sensitive condition, and increased the size of the vibrator plates to 6 feet by 6 feet. One day in February 1896, after some failures, he was able to activate the detector at a friend's house about half a mile from the Cavendish Laboratory.

In January 1896, in his letter to his family, Rutherford wrote: "If I could get an appreciable effect at ten miles, I would probably be able to make a considerable amount of money out of it, for it would be one of great service to connect lighthouses and lightships to the shore so that signals could be sent at any time." J. J. Thomson too sensed this possibility. In February 1896, he wrote to William Thomson (then Lord Kelvin): "A pupil working in the laboratory . . . was able to send signals by the electric waves to the rooms of a friend who lives more than half a mile away though the electric waves had to go through a very thickly populated part of Cambridge. I should think an instrument of this kind of might prove serviceable for communicating between a light house and the shore in bad weather or perhaps even between ships." In April, J. J. Thomson reminded Kelvin of Rutherford's work again: "Rutherford . . . is still working away perfecting the apparatus and applying it to solve many interesting questions about these waves. His work will I think be of great importance." There is no surviving reply from Kelvin, but J. J. Thomson's biographer (the fourth Lord Rayleigh) tells us that Kelvin's estimate that such a business would require an initial capital expenditure of £100,000 discouraged both J. J. Thomson

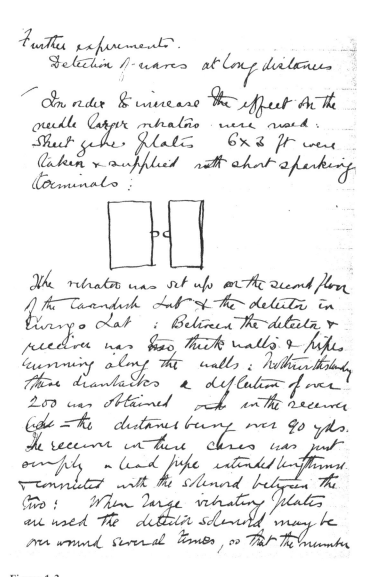

Figure 1.3
"Further refinements. Detection of waves at long distances" in Rutherford's notebook (Cambridge University Library).

and Rutherford. In the spring of 1896, Rutherford moved on to the topic of Röntgen rays, in which he was quickly absorbed. Rutherford summed up the results of his research on Hertzian waves, read a paper on this before the Royal Society in June, and gave the same presentation before the Cambridge Natural Science Club in July and before the British Association for the Advancement of Science in September. At the latter meeting, however, he heard the news that an Italian named Marconi had succeeded in transmitting a mile and a half.[19]

Henry B. Jackson was an engineer associated with the torpedo school of Royal Navy.[20] As the captain of HMS *Defiance* (a ship used by the navy's torpedo school) in 1895–96, he had much interest in secret communication. As a matter of fact, he had come up with the idea of using Hertzian waves for communication as early as 1891. The motivation for his 1895 research on wireless telegraphy came from Jagadis Bose's experiment on the optical properties of Hertzian waves in the same year. Bose had invented a stable spring coherer which *The Electrician* had appraised as a "workmanlike form of coherer" that might be suited as a receiver on board to detect signals from "electromagnetic 'light'-houses." Bose had read his articles before the Royal Society, and they had been published in *The Electrician*.[21] Inspired by Bose, Jackson bought a 1-inch induction coil and a spring coherer and began experimenting in December 1895. The first results "were not very encouraging, . . . but some results were obtained." The spring coherer was tapped by a finger. After this first trial, Jackson read Lodge and Hertz and had "extracted much useful information." In March 1896, he employed a 2-inch induction coil. In July, he tried a glass tube coherer. In August, he succeeded in "obtaining intelligible Morse signals through the length of the ship through all the wooden bulkheads," the distance being about 100 yards. His receiver was an electric bell that was also used for tapping the coherer. After this success, Jackson began experimenting with a Morse sounder.[22]

In September 1896, Jackson was called on to inspect the instruments used in Marconi's Post Office and Salisbury Plain demonstrations (discussed in chapter 3 below). His report on Marconi's demonstrations to the Navy reads as follows:

The inventor claims by its means to transmit electric signals to a distance without connecting wires, and also by a further development of this, to work the helm of a ship, boats, or torpedos in motion. . . . I have personally, for the last six months, been experimenting in the same direction as Mr. Marconi, with apparatus differing but little in detail or in the main principles involved; and with results lately similar, though not so good as his, owing partly to the want of powerful transmitting apparatus, and the use of less sensitive receiving apparatus at my disposal on board, and partly, perhaps principally, to the materials and construction of the receiver, which is one of the most important parts of the apparatus, and the exact construction of which the inventor will not disclose until patented, and constitute a most important detail, and one which has involved much time and trouble in bringing it to its present perfection.[23]

In November 1896, Jackson, who had become a close friend of Marconi, used a printer to record signals. In April 1897 he succeeded in transmitting signals over 2 miles, but the antenna he used was of Marconi's design. In May he reported:

Comparing my experiments with those of Mr. Marconi, I would observe that before I heard of his results, I had succeeded with the instruments at my disposal in transmitting Morse signals with my apparatus about 100 yards, which I gradually increased to one-third of a mile, but could not improve upon till I obtained a more powerful induction coil last month, with which I have obtained my present results, using Marconi's system of wires insulated in the air attached to transmitter and receiver. . . . With this exception, the details of my apparatus, which so closely resemble his, have been worked out quite.[24]

Marconi and Telegraphy before Wireless

Hertzian devices were widely available to Victorian scientists and engineers. However, optical analogies, rather than telegraphic ones, dominated. Since light and electromagnetic waves were of the same kind, an almost perfect analogy between light receptors and wave detectors was sought. The use of Hertzian waves was foreseen as a practical means of ship-to-lighthouse signaling on foggy days. Such signaling, which had been an important scientific and technological goal for many British scientists and engineers and for the government, constituted an obvious social need for wireless communication. In this sense, the optical analogy acted as a resource.

However, the optical analogy acted as a constraint too. The flasher used for lighthouse signaling became a technological prototype for Hertzian wireless communication. Trotter, Threlfall, and Rutherford, none of whom had any background in telegraphy, worked under this constraint. Further, optical analogies led scientists to shorten the wavelength of waves. Everyone other than Rutherford used short waves, or thought that short waves should be exploited for practical purposes. Trotter proposed using a 1-millimeter wave; Crookes talked about reducing the wavelength; Jackson's initial device seems to have generated waves in the range of a few centimeters, for it was essentially Bose's microwave optical instrument (figure 1.4). Only Rutherford seemed to sense that a powerful generator with a longer spark was the key to successful long-distance transmission. To increase power, he used large condenser plates (6 feet by 6 feet) of huge capacitance, which, in effect, increased the wavelength. But Rutherford changed his research topic before he could design a workable communication device. Unlike Rutherford, Jackson tried to transmit Morse-coded messages, but the limited transmitting distance (about 100 yards) was a serious hindrance. This deficiency was not confined to Jackson. Even after hearing the news of Marconi's success in sending wireless messages by means of Hertzian waves, the German physicist Adolf Slaby, who had replicated Marconi's telegraphic devices in January 1897, remarked that he "had not been able to telegraph more than one hundred meters through the air" (Slaby 1898, p. 870).

These considerations illuminate Marconi's success from a different angle. Marconi was born in 1874. His father, Giuseppe Marconi, was a rich Italian gentleman who had a small estate in Pontecchio, near Bologna. His mother, Annie Jameson, was from an influential family in Ireland. His mother's Irish connection later turned out to be valuable when Marconi started his business in Britain. His family lived for a time in the Villa Grifone in Pontecchio.

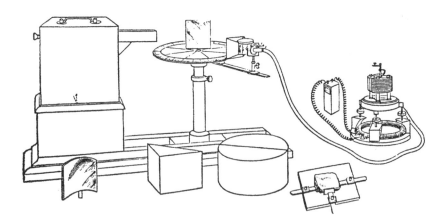

Figure 1.4
The arrangement of devices for Bose's short-wave optical research. Source: Bose 1895a.

In winter they often stayed in Florence or in Leghorn (Livorno), where Marconi irregularly attended elementary school. By the age of 14, Marconi had become interested in invention and in scientific topics. Since schools in Leghorn did not teach advanced science, he took private lessons on advanced electrical science from Vincenzo Rosa, a professor of physics at the Liceo Niccolini, and performed some physics experiments in Rosa's laboratory. Later, after his family stopped wintering in Leghorn, Marconi attended the lectures of Professor Augusto Righi at the University of Bologna.[25]

As a student of the Technical Institute in Leghorn, Marconi learned about cable telegraphy from an old telegrapher (D. Marconi 1982, pp. 16–17). He was inspired to work on telegraphy with Hertzian waves by Augusto Righi's obituary of Hertz in the journal *Nuovo Cimento*, which he read in the summer of 1894. During the winter of 1894–95, Marconi spent most of his time in an attic laboratory at his family's villa in Pontecchio, becoming familiar with such Hertzian devices as spark-gap oscillators, condensers, resonators, coherers, and induction coils. In the summer of 1895, Marconi began experimenting with Hertzian wave telegraphy. At first even simply receiving Hertzian waves proved difficult, but the situation gradually improved with Marconi's experimental skills. Around this time, he visited Righi's laboratory at the University of Bologna, where centimeter-range waves were being investigated in quasi-optical experiments. Marconi modified Righi's high-frequency quasi-optical instruments for telegraphy. For example, he added a small dipole radiator to each side of the balls in Righi's four-ball spark transmitter, and this greatly lengthened the waves.[26]

Marconi had difficulties with his receiver. He started with a Branly-tube coherer and a galvanometer. At first he tapped the coherer by hand, as had been done in Henry Jackson's later experiments. However, he found the Branly coherer unstable: it "would act at thirty feet from the transmitter, [but] at other times it would not act even when brought as close as three or four feet." To receive telegraphic signals, he realized, "something more reliable than the Branly tube" was needed. The original Branly tube coherer was long and bulky. In Marconi's hands, it evolved into a small (36 mm long, 3 mm in diameter) evacuated tube. After trying 300–400 kinds of metallic filings in his coherer, Marconi found that a mixture of nickel and silver filings (with the addition of mercury drops) produced the best sensitivity. Marconi's new coherer proved to be much more reliable than the Branly tube, but tapping by hand still made it unstable. To correct that, he designed an electric tapper with a small hammer and an electromagnet. Similar to an electric bell, the tapper was designed to be activated by the coherer current. Marconi supposed that the tapper would be very stable, since it remained electrically dead when there were no incoming waves. However, the coherer current (from a single battery) was not strong enough to activate the electric tapper. Fortunately, this problem had already been solved in cable telegraphy by the use of a relay device to augment the current. Marconi connected a telegraphic relay in series to the coherer and made the relay, which was activated by the change in conductivity of the coherer, activate the tapper. He then replaced the galvanometer with a Morse inker. However, the electric tapper and the relay caused another unexpected problem: a local electromagnetic disturbance from these devices sometimes activated the coherer, producing unnecessarily long dashes and dots. This was fixed by connecting a high-resistance shunt circuit to the tapper and relay parts of the circuit. Finally, Marconi attached condenser plates to his coherer to make his receiver respond more readily to incoming waves.[27]

Everything, in principle, was ready for wireless telegraphy. In August 1895, Marconi conducted a series of experiments in the field to determine how far his wireless signals could travel, since the transmitting distance had appeared to exceed the dimensions of his laboratory. He took both his transmitter and receiver out of the laboratory. The transmission distance was not more than 150 feet. Marconi thought that received waves were being dissipated by the battery circuit of the receiver. To prevent this, he inserted a high inductance between the battery circuit and the coherer, to prevent any received high-frequency oscillation from dissipating, while leaving the flow of the battery current intact. With this improvement, he succeeded in transmitting a distance of ⅛ mile.[28]

Marconi enlarged the size of the plates attached to the transmitter and the receiver up to 6 feet by 6 feet, coincidentally the same size Rutherford was using. Also coincidentally, the transmitting distance Marconi achieved with this modification, in August 1895, was ½ mile, the same distance Rutherford would achieve in February 1896 and the same distance that Lodge had predicted in 1894 as the maximum transmitting distance with Hertzian waves. It was at this point that Marconi made his most significant breakthrough. He connected one end of the plate of the receiver, and one end of the transmitter, to the earth (figure 1.5). With this modification, he could transmit

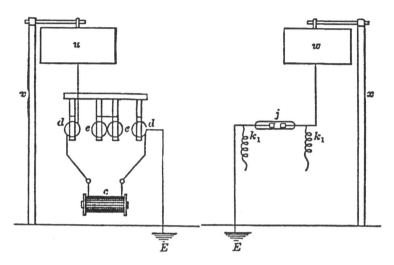

Figure 1.5
Marconi's grounded antenna. Source: Marconi, British patent specification 12,039 (1896).

messages a mile, and eventually up to 2 miles over hills. The higher the transmitter pole, the further the signals would go. The first practical wireless telegraphy was born. In his notebook, Marconi recorded the following:

A coil giving a three inch spark was used. With cubes of tin $30 \times 30 \times 30$ centimeters, at the transmitter and receiver, on poles 2 meters high signals were obtained at 30 meters from the transmitter, with the same cubes on poles 4 meters high signals obtained at 100 meters, and with the same cubes at a height of 8 meters (other conditions being equal) morse signals were easily obtained at 400 meters. With large cubes $100 \times 100 \times 100$ centimeters, having a surface of 6 square meters, fixed at a height of 8 meters, reliable signals were obtained at 2,400 meters all round, equal to 1 mile and 1/2.[29]

What led Marconi to ground his apparatus and to elevate his poles? The idea of vertical elevation itself was not new. Hertz had once set his transmitter vertically, and Minchin had used a long wire in his receiver in an effort to collect electromagnetic waves. But Marconi was the first to ground both the transmitter and the receiver.[30] Marconi got this idea from cable telegraphy, for which a "good earth" was essential. It should be remembered that Marconi's original aim was to invent "telegraphy" with Hertzian waves. He gradually perfected the symmetry between wired and wireless telegraphy by employing a telegraphic key, the Morse-coded message system, a Morse inker, and an ordinary telegraphic relay in his research on

Hertzian waves. His earth connection in his experiments with wireless telegraphy is analogous to an "earth-return" in wired telegraphy. The emphasis on telegraphy separates Marconi from the other physicists and engineers who were working on Hertzian waves; all the others were preoccupied with optical analogies.

Understanding Marconi's Originality

Between 1888 and 1896, Hertzian waves were transformed from laboratory entities into useful technology. This transformation was neither smooth nor straightforward. Hertzian devices emerged from experimental physics. Lodge, FitzGerald, Trouton, Crookes, Rutherford, J. J. Thomson, Threlfall, Minchin, and other physicists manipulated Hertzian waves in the laboratory and explained their properties through an analogy to light, but they only speculated about utilizing the waves for communication. However, the physicists and their experiments inspired engineers. For example, Righi inspired Marconi through his obituary of Hertz and through his experiments. Bose, who read a series of articles on Hertzian optics in London in 1895, inspired Jackson to start experimenting with Hertzian devices for purposes of communication between navy ships. In contrast, telegraphists and telegraphic engineers had little interest in Hertzian waves and devices. William Preece, the best-known telegraphic engineer in Britain, who was keen on establishing workable wireless communication between a lightship and a lighthouse, constantly tried inductive, as well as earth-conductive, wireless telegraphy in the 1880s and the early 1890s, but he never thought about using Hertzian waves for wireless communication (Preece 1893, 1896c).[31]

The physicists, however, were preoccupied with optics. Quasi-optical questions, optical devices (such as polarizers and diffraction gratings), and short waves were standard features of physical experiments with Hertzian waves. There were also theoretical and practical limitations to the scope of these experiments. Theoretically, half a mile was the maximum transmitting distance with Hertzian waves. Practically, no one but Rutherford even came close to that. Galvanometers, in connection with Branly's tube, Lodge's coherer, or Rutherford's magnetic needle, were invariably used as receivers. Those who attempted to use a Morse inker with Branly's tube or Lodge's coherer found that it was not activated by either.[32] A telegraphic relay could have been employed to magnify the effect of the coherer current, but it generated local sparks that were hard to eliminate.

Such theoretical and practical barriers were not impossible to overcome. After several months of hard work, Marconi had devised a sensitive and stable coherer, invented a stable tapper, increased the efficiency of the induction coil by perfecting its insulation, connected the Morse inker and the telegraphic relay to the coherer in the receiver circuit, and inserted a shunt resistance and an inductance into the receiver circuit to prevent, respectively, local sparks and easy dissipation of received signals. Although most of these components had been invented by others, they were unstable or unconnected with one another before Marconi. Marconi's inventions, modifications, and improvements fit into a small box, at that time dubbed Marconi's "secret-box" or "black-box." When Marconi "opened" this "black-box" by publicizing his first patent in 1897, people were amazed and intrigued by its simplicity. The solutions appeared so simple and so obvious that many began to wonder why no one else had come up with them.

There was one element of "non-obviousness" in Marconi's solutions: his grounding of one pole of the transmitter and one pole of the receiver. With this advance, Marconi became the first to transmit messages more than half a mile through buildings and across hills.

I have suggested that the source of Marconi's innovative antenna was his analogy between wired and wireless telegraphy. By grounding one pole, he was able to use the earth (or the "power" of the earth) for communication, as telegraphers had done in cable telegraphy. Later, he came to understand that grounding coupled with an antenna multiplied the capacitance of his transmitter, which generated long waves. Over a relatively short distance, these long waves were propagated along the surface of the earth, as if the earth was guiding them. This surface wave was effective only over a short distance. However, Marconi believed that with enough power such a wave could travel along the earth's surface to transmit signals across the Atlantic. When he successfully transmitted signals across the Atlantic in 1901, he attributed his success to earth-guided waves. The old telegraphers' "good earth" proved to be essential to Marconi's understanding of the new wireless telegraphy.[33]

2

Inventing the Invention of Wireless Telegraphy: Marconi versus Lodge

The point is which of the two was the first to send a wireless telegram? Was it Lodge in 1894 or Marconi in 1896?
—Silvanus P. Thompson, *Times* (London), July 15, 1902

As was discussed in chapter 1, several engineers and physicists worked with Hertzian waves and tried in various ways to use them. None among them succeeded in producing a practical system, in part because of their specific understandings of Hertzian waves as optic-like entities. However, one British physicist did fight with Marconi about priority and about patents on wireless telegraphy: Oliver Lodge. Indeed, Lodge's 1894 lectures, particularly one that he gave at an August meeting of the British Association for the Advancement of Science at Oxford, were widely seen in Britain as establishing his claim to have been the first to demonstrate wireless telegraphy, and several historians accepted those claims.[1] Since Marconi did not use Hertzian waves for signaling until 1895, Marconi's priority could not be sustained if Lodge's claim were correct.[2]

In this chapter I will take up the priority dispute between Marconi and Lodge in regard to the invention of wireless telegraphy. My analysis will show that any claim for Lodge's priority is incorrect. After considering evidence to the contrary, I will turn to what was later claimed to be Lodge's demonstration of wireless telegraphy in 1894. This, it will be seen, had nothing to do with telegraphy, with alphabetic signals, or with dots and dashes. I will then turn to the impact of Marconi and his British patent on British Maxwellian physicists, particularly Lodge, Thompson, FitzGerald, and Fleming.[3] The transformation of Hertzian laboratory apparatus into commercial wireless telegraphy was, as will also be seen, clearly accomplished by Marconi, an Italian "practician," and this created a certain disharmony between theory and practice. Moreover, Marconi's patent appeared to be so strong that it threatened to monopolize Hertzian waves and thereby impact the British national interest. The image of Lodge as the inventor of wireless telegraphy was deliberately constructed by his friends and by Lodge himself to counter the effects of these social circumstances.[4]

Fleming's Marconi Memorial Lecture of 1937

The priority dispute between Lodge and Marconi has been discussed extensively. The historian of technology Hugh Aitken and others who support Lodge's priority have proposed two pieces of textual evidence. The first is a short article in *The Electrician* according to which "both at Oxford [in August 1894] and at the Royal Institution [in June 1894], Dr. Lodge described and exhibited publicly in operation a combination of sending and receiving apparatus constituting a system of telegraphy substantially the same as that now claimed in the patent we have referred to [Marconi's patent 12039 of 1896]."[5] The second is the Marconi Memorial Lecture of 1937, in which John Ambrose Fleming (Marconi's scientific advisor) said that Marconi was "not the first person to transmit alphabetic signals by electromagnetic waves" and admitted Lodge's priority (Fleming 1937, p. 46; cited in Aitken 1976, p. 123):

[Lodge] was able to transmit a dot or a dash signal and by suitable combinations to send any letter of the alphabet on the Morse code and consequently intelligible messages. He had also on his table a Morse inker (so he tells me), and could have used it with a sensitive relay to print down the signals, but as he wished the audience to see the actual signals he preferred to use the mirror galvanometer. It is, therefore, unquestionable that on the occasion of his Oxford lecture in September, 1894, Lodge exhibited electric wave telegraphy over a short distance.[6]

Before 1896, when Marconi arrived in England, Fleming and Lodge had been close friends, having studied together in their youth in the laboratory of the chemist Edward Frankland in South Kensington. Their relationship deteriorated rapidly, however, after 1899, when Fleming became a scientific advisor to Marconi's Wireless Telegraph and Signal Company. Until 1937, Fleming would not admit Lodge's priority. For example, in 1906 he reiterated that "no mention of the application of these waves to telegraphy was made" at the 1894 Oxford meeting (Fleming 1906a, p. 424; Aitken 1976, p. 120). Only after Marconi's death, it seems, did Fleming decide to tell a different story.[7] Aitken (1976, p. 122) comments that "Fleming's memory . . . was capable of improvement with the passage of time, or perhaps as commercial and scientific rivalries receded into the past."

But Fleming was not present at the August 1894 meeting of the British Association for the Advancement of Science. Fleming's source was not his own memory but Lodge's remark. This is clearly evident from three letters exchanged by Fleming and Lodge in 1937. Before his Marconi Memorial Lecture, given in November, Fleming wrote to Lodge:

I have been asked by the Council of the Royal Society of Arts to give next November 10th a Memorial Lecture on the "Work of Marconi." . . . One of the facts I am anxious to learn about is whether in your lecture to the *British Association at Oxford Meeting in 1894* you used a telegraphic relay in series with your coherer to print on Morse Inker tape *dot* and *dash* signals? In his little book on "Wireless" [Eccles 1933] Dr. Eccles gives on page 54 a diagram of the apparatus he says you employed at Oxford in 1894 [see figure 2.1] . . . I was present in June 1894 at your famous lecture at the Royal Institution on "The Work of Hertz" and remember well your experiments with your coherer. But to the best of my recollection there was no direct

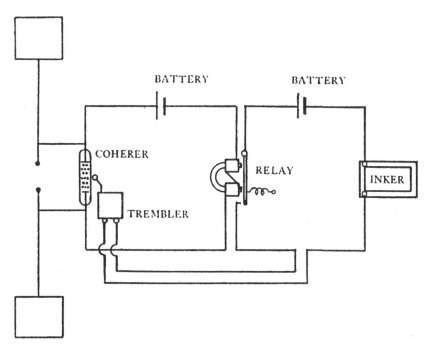

Figure 2.1
W. H. Eccles's diagram of Oliver Lodge's 1894 receiver. Source: Eccles 1933, p. 54.

reference to "telegraphy" in that lecture. I was not present at the B.A. meeting at Oxford, but . . . it is very important to know from you whether at Oxford in *1894* you exhibited a Hertz oscillator connected with coherer and used a telegraphic relay in connection with it and morse inker and showed the transmission and printing of *dots* and *dash* signals over any short distance.[8]

Lodge replied that at Oxford he had actually used telegraphic instruments and transmitted alphabetic signals, i.e., dots and dashes:

You are perfectly right that in 1894 at the Royal Institution I did not refer to telegraphy. But, stimulated by Muirhead, who had close connection with telegraphy and cables, I did at Oxford demonstrate actual telegraphy. I had a Morse instrument there, but it was not convenient for the large audience in the Museum theatre, and therefore I used as a receiver a Thomson marine signalling device supplied by Muirhead's firm for that purpose, though I had a morse instrument on the table which I could have used instead. But the deflections of the spot of light were plainly visible to the audience, and gave quick and prolonged response corresponding to the dots and dashes according to the manipulation of the key at the distant end.[9]

Fleming's reply to Lodge, which foretold the content of his lecture, shows that he entirely accepted Lodge's claims:

What you tell me about your Oxford lecture in 1894 is very valuable and important. It is quite clear that in 1894 you could send and receive *alphabetic signals* in Morse Code by Electric Waves and did send them 180 feet or so. Marconi's idea that he was the first to do that is invalid. . . . Marconi was always determined to claim everything for himself. His conduct to me about the first transatlantic transmission was very ungenerous. . . . However, these things get known in time and justice is done.[10]

As is evident from the last letter, Fleming had been hurt by Marconi's attitude toward his employees. Fleming's resignation as scientific advisor to the Marconi Company in 1931 and Marconi's death in 1937 may have influenced Fleming to be more sympathetic to Lodge. He may have felt that "things get known in time and justice is done."[11] But this could not have improved his memory of something he had never experienced. It was Lodge who informed Fleming about the Oxford meeting. Therefore, Fleming's lecture in 1937 cannot be regarded as conclusive.

For purposes of analysis, I will divide Lodge's claim into two parts: (1) that Lodge actually sent telegraphic signals (i.e., dots and dashes) during the August 1894 meeting of the British Association for the Advancement of Science in Oxford and (2) that Lodge had a Morse instrument there but, owing to the size of the audience, used a mirror galvanometer to show the signals. It will become evident that neither assertion is correct.

Hertzian Waves and "Lodgian" Practice in 1894

The ambitious Maxwellian Oliver Lodge, a professor of physics at University College Liverpool, worked on various characteristics of Hertzian waves between 1888 and 1894. The links between optics and electromagnetism particularly attracted him. The subject was faithfully Maxwellian, as it had a root in Maxwell's doctrine that light and electromagnetic waves were the same. It was also truly Lodgian "imperial science," as it led electrical science to the conquest of other fields—in this case, optics and physiology. The subject was divided into two parts: (1) physical investigation of the quasi-optical property of electromagnetic waves—that is, reflection, refraction, and polarization of the electromagnetic waves in air, in other media, and in some cases along the wires—and (2) physiological investigation of the mechanism of the perception of light (color, intensity, and so on) by the human eye.[12]

With his experiments, Lodge made two important advances. First, he constructed a radiator that generated waves with wavelengths of several inches. Hertz had once used a wavelength of 66 cm, but that was still too long for most optical experiments. Because of the difficulty of decreasing the wavelength with Hertz's dipole radiator, Lodge turned to a spherical radiator. In 1890, Lodge used three 12-cm balls and obtained 17-cm waves, "the shortest yet dealt with" (Lodge 1890a). Lodge then went further along this line of development and devised two more spherical radiators, which he exhibited during a Friday evening lecture titled "Work of Hertz" at the Royal Institution in June 1894. Lodge's second line of research concerned detectors. The Hertzian wave was at first detected by a small spark-gap resonator. But this spark-gap resonator was not suitable for Lodge's physiological research. For example, nothing in the spark-gap detector corresponded to different color perceptions in human eyes. Lodge therefore concentrated on the construction of an "electric eye." In 1890, his assistant Edward Robinson constructed a "gradated receiver," and Lodge tried "a series of long cylinders" of various diameters. The principle of both detectors was to respond to a specific radiation, forming "an electric eye with a definite range of colour sensation." In 1891 Lodge exhibited an electric eye of the Robinson type at a meeting of the London-based Physical Society. That device had "strips of tin foil of different lengths attached to a glass plate, and spark gaps at each end which separate them from other pieces of foil" (Lodge 1890a, 1891a). As was noted in chapter 1, Lodge also constructed a spring point-contact detector, which he called a coherer.

Lodge therefore had two new detectors: his coherer and Branly's tube. (Initially, Lodge called only his single-point detector a coherer, but that name soon came to designate both types.) Lodge's coherer and Branly's tube were connected in series to a battery and a galvanometer. Under this condition, either device acted like an on-off switch: before a Hertzian wave struck it, its resistance was very high, as if the switch were off, but when a Hertzian wave struck it, its resistance decreased, as if the switch had been turned on. This action made current flow from a battery, and the current could be detected by a galvanometer. The two detectors, however, differed in sensitivity. At Liverpool on April 17, 1894, Lodge found that the filing tube could detect radiation emitted from 40 yards away. On the other hand, Lodge's biographer Peter Rowlands (1990, pp. 116–117) notes that "a sender in Zoology Theatre affected the coherer in Physics Theatre perceptibly."[13] Though more sensitive, Lodge's coherer was less stable than the filing tube. In addition, Branly's tube had a crude metrical character: its decrease in resistance seemed roughly proportional to the intensity of the Hertzian waves. This resembled the human eye's perception of the light of different intensity. For physiological experiments, therefore, Branly's tube was more suitable than Lodge's single-point contact coherer.

Hertz died at the age of 36 on January 1, 1894, and on June 1 Lodge delivered a Hertz Memorial Lecture at the Royal Institution's Friday Lecture. Here Lodge spoke on Hertz's life and work, exhibited Hertz's and his own radiators and detectors, then performed several experiments.[14] The demonstrations were divided into a physical part and a physiological part. In the physical part Lodge demonstrated reflection, refraction, and polarization of the Hertzian waves. For this purpose he used his spherical radiator enclosed in a metallic box and a Branly tube in a copper "hat" as a detector, and a mirror galvanometer as a signal indicator (figure 2.2). In the physiological part he explained the functioning of the human eye, using the analogy of the coherer: "When light falls upon the retina, these gaps become more or less conducting, and the nerves are stimulated." (Lodge 1894b, p. 137) Lodge also tried an outdoor experiment in which the receiver was in the theater and the transmitter was in the library of the Royal Institution, separated across a distance of 40 yards by three rooms and stairs.

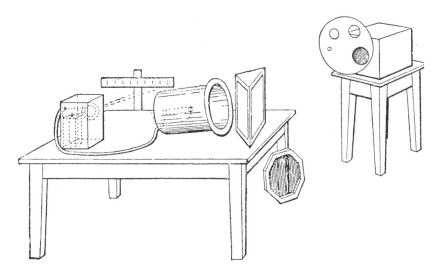

Figure 2.2
Oliver Lodge's quasi-optical experiment with Hertzian waves at the Royal Institution in June 1894, with a spherical radiator in a metallic box and a Branly tube in a copper "hat." Notice the mirror galvanometer. Source: *The Electrician* 33, June 1, 1894, p. 205.

After detecting electromagnetic waves, the coherer had to be mechanically vibrated or "tapped" to make it ready for the next wavetrains. To what in the human eye did this tapping correspond? Lodge assumed that in the eye "the tapping back is done automatically by the tissues, so that it is always ready for a new impression." To demonstrate this automatic tapping in the human eye, Lodge prepared an electric bell, which was mounted on the same board as the filing tube. By constantly vibrating itself, and thus constantly shaking the table and the coherer on it, the bell always made the coherer ready to detect new waves (Lodge 1894b, p. 137).[15]

The abstracts of Lodge's Friday Lecture in *Nature* and in *The Electrician* were read worldwide. The demonstrations, however, were rather unsuccessful. *The Electrician* noted that "the experiments were performed under very unfavourable conditions."[16] Moreover, the "lack of enthusiasm" in Lodge's lecture was contrasted with the success of Nikola Tesla's 1893 lecture, during which "the weird waving of glowing tubes in the suitably darkened room" impressed everyone. What Lodge lacked was a "theatrical effect" or a "scenic setting." Neither the sound of the spark nor the "moderate galvanometer" connected to the coherer was theatrical. The "very lively" galvanometer was particularly tricky; the swing of the needle was not stable, even when there were no waves. *The Electrician* suggested that a more effective galvanometer (e.g., one of the "dead-beat" type[17]) be used.

No detailed descriptions of the galvanometer used by Lodge survive, but from the abstract and the printed figure in *The Electrician* it is evident that Lodge used a mirror galvanometer.[18] From a comment in *The Electrician*, it is evident that it was not of the dead-beat type. From other pieces of evidence, we know that Lodge had not paid much attention to the galvanometer. Before the invention of the coherer, for example, George FitzGerald had constructed an extremely sensitive galvanometer in order to demonstrate the detection of waves to an audience. This instrument would have had to detect the disturbance of electric equilibrium caused by a tiny spark (FitzGerald 1890). The coherer, an on-off switch, made such a sensitive galvanometer unnecessary, because the galvanometer had to detect only a relatively large current from a battery—a current triggered by the action of the coherer. As Lodge noted, "a rough galvanometer" was therefore sufficient.[19]

But why was the galvanometer troublesome at the crucial moment? Lodge suspected that the source of the trouble was a "jerk current" in the electric bell used for the automatic tapping. Such a current would certainly influence the adjacent coherer electrically. The jerk current "produces one effect, and a mechanical vibration . . . produces an opposite effect; hence the spot of light can hardly keep still." Lodge (1894b, p. 137) knew that a "clockwork" that did not use an electric current "might do better" than the electric bell.

In the letter to Fleming cited above, Lodge emphasized that in his Oxford demonstration he had used a dead-beat Thomson (Kelvin) marine galvanometer borrowed from Alexander Muirhead's firm. In 1900 Lodge stated: "Dr. Alexander Muirhead foresaw the telegraphic importance of this method of signaling immediately after hearing the author's lecture on June 1st, 1894, and arranged a siphon recorder for the purpose." (Lodge 1900, p. 45; Lodge 1921–22) In a much-quoted 1914 letter to one of his friends, Lodge wrote: "It was at the first of [the Royal Institution's Friday Lectures] that my friend Alexander Muirhead conceived the telegraphic applications which ultimately led to the foundation of The Lodge-Muirhead Syndicate."[20] Elsewhere, Lodge recalled that the galvanometer at Oxford "responded to signals sharply, in a dead-beat manner, without confusing oscillations" (Lodge 1926, pp. 265–266).[21] This Muirhead connection would make Lodge's telegraphic trial at Oxford, performed only 2 months after his obviously non-telegraphic experiment at the Royal Institution, both feasible and timely.

It is true that Muirhead constructed a delicate siphon recorder for a wireless detector during the late 1890s and the early 1900s, and that Lodge and Muirhead, who began to file for patents on wireless telegraphy in 1897, formed the Lodge-Muirhead Syndicate in 1901. Nevertheless, the central assertion in regard to a Muirhead connection—that Muirhead lent Lodge a dead-beat Thomson marine galvanometer after being inspired by Lodge's June lecture—is dubious. According to the recollection of Muirhead's wife, it was Lodge's Oxford lecture, not the Royal Institution lecture, that inspired Muirhead to think about wireless telegraphy.[22] One of Lodge's biographers doubts that Lodge used a Thomson marine dead-beat galvanometer borrowed from Muirhead at Oxford (Rowlands 1990, p. 148, note 30).[23] But Lodge's use of a Thomson marine galvanometer at Oxford is, I think, highly plausible, not because Muirhead had been inspired by Lodge's June lecture, but because Lodge had borrowed from Muirhead such a device at various times since the late 1880s.[24] In addition, as we have seen, Lodge had an urgent reason to use a dead-beat galvanometer. He had experienced serious trouble with his "lively" galvanometer in the June lecture, and *The Electrician* had recommended the employment of a dead-beat galvanometer for future success. These factors might be the real motivations for Lodge's use of a Thomson marine galvanometer at Oxford, if it was actually used there.

Why did Lodge perform his outdoor experiment at the Royal Institution? Not, evidently, to determine the maximum transmitting distance, or to show the wave's penetrability of walls. The real purpose had to do with physiological concerns. With a metrical Branly tube and an electric bell, Lodge wanted to show that the coherer could discern Hertzian waves of various intensities, just as the human eye could. The easiest way to vary the intensity of waves was to adjust the distance between the transmitter and the receiver. Owing to the theory of Horace Lamb and J. J. Thomson, Lodge's use of a 6-inch spherical radiator (placed in the library of the Royal Institution) made it easy to estimate the wavelength of the wave produced from the spherical radiator: with a 6-inch radiator, the wavelength was about 8 or 9 inches. Furthermore, owing to FitzGerald's theory (1883), it was known that the energy of radiation at a distance, other things being equal, is inversely proportional to the fourth power of the wavelength. That is, the shorter the wavelength, the more the energy of radiation, and thus the higher the possibility of being detected at a distance. The belief that short waves could travel farther than long waves was strongly inscribed in the mind of Lodge (Lodge 1931, p. 165; Lamb 1883; J. J. Thomson 1883–84). But even with the short waves, Lodge (1894b, pp. 135–137) esti- mated that half a mile was perhaps the limit of sensitivity.

Was Lodge's first outdoor experiment successful? Lodge and his friends later repeatedly called it a great success. The correct answer, however, is yes and no: no because he failed to detect the wave with a metrical filing tube, yes because he detected it with his sensitive coherer. Lodge's manuscript confirms this judgment:

The spherical radiator . . . though it could excite the filings tube . . . when 60 yards away in the open air, . . . could not excite it perceptibly when screened off by so many walls and metal surfaces as exist between the Library and Theatre of the Royal Institution. It could, however, still easily excite the coherer, which is immensely more sensitive, and also more troublesome and occasionally capricious than is a tube of iron filings.[25]

On August 14, 1894, at a joint session of the Physics and Physiology sec- tions of the British Association for the Advancement of Science, Lodge delivered two lectures and demonstrations on Hertzian waves in a theater at Oxford University. The first lecture was titled "Experiments Illustrating Clerk Maxwell's Theory of Light"; the second was titled "An Electrical Theory of Vision." In a sense, he split the previous Friday's lecture into two. In the first lecture, Lodge used a spherical radiator and a copper hat to con- centrate the radiation. As before, Branly's tube and Lodge's coherer were used as detectors, and most of the demonstrations employed Branly's device. Refraction and reflection of Hertzian waves were demonstrated with lenses, gold papers, the human body, paraffin prisms, and a slab of wood. Polarization was demonstrated with a copper wire polarizer. Splitting of the polarized ray into the two elliptically polarized rays was also demon- strated. The phenomena were "very beautifully, very carefully and very con- vincingly demonstrated," and "the audience . . . repeatedly showed its warm appreciation of Prof. Lodge's beautiful experiments."[26] In the second lecture, Lodge proposed that the coherer circuit "may be taken as an analogous, and may, *ex hypothesi*, be an enlarged model of the mechanism of vision." According to this hypothesis, "the retinal elements constitute an imperfect conductor, and . . . the light waves would cause a sudden diminu-tion in the resistance of the elements." Yet, once struck by the wave, the coherer "has a tendency to persist in its lessened resistance" and therefore requires tapping "to jerk the coherer contact back to its normal state of badness." For this tapping, Lodge used "a sort of clockwork apparatus which automatically produces the tap every tenth of a second" to show that "for a continuous radiation the coherer showed continuous indications, which died away when the radiation ceased."[27]

Where was the transmitter in the physiological experiments? This has never been examined critically. In 1898, Lodge's close friend Silvanus Thompson, a professor of applied physics and electrical engineering at Finsbury Technical College, reported that the radiator had been in the Clarendon Laboratory, at a distance of 200 yards (S. P. Thompson 1898, p. 458). Thompson's statement may be erroneous, however. Lodge's vari-ous recollections contain no mentions of the Clarendon Laboratory. We have only his remark that "in both cases, signaling was easily carried on from a distance through walls and other obstacles, an emitter being outside and a galvanometer detector inside the room" (1897, p. 90) and his remark that "this [sending] apparatus was in another room" (1932, p. 231). Contrary to Lodge's and Thompson's remarks, the four sources on which I have relied say nothing about the outdoor trial at all. In view of this evi-dence, and of Lodge's previous trouble with the outdoor experiment at the Royal Institution, it may be the case that the distance traversed by Hertzian waves in the Oxford lectures was fairly small.

There is not the slightest hint of telegraphic signals, or of dots and dashes. With his improved automatic tapper, Lodge showed the persistence of vision and mere sensation of light, which corresponded to the continuous and short indication of the galvanometer. But that was far from dots and dashes for alphabetic signals. From beginning to end, the lecture was entirely "Lodgian." Its purpose was to investigate the relation between optics and electromagnetism, between light and electromagnetic waves, and between optical receptors and electromagnetic ones. After the lecture, despite Muirhead's and Rayleigh's suggestions, Lodge did not pursue this subject further. He soon busied himself with ether experiments, x rays, and psychic research.

Next let us examine Lodge's second argument about "a Morse instrument," mentioned in his letter to Fleming. Fleming thought that this instrument must be a Morse inker. But it was not. Ironically, the first evidence (a short article in *The Electrician*) reveals its nature. It listed five instruments used in Lodge's Oxford demonstrations, and one of them is "Morse instrument to shake the filings."[28] Lodge's Morse instrument was nothing but a clockwork or an automatic tapper that he used for tapping the coherer in place of the troublesome electric bell. To be sure, the Morse instrument that Lodge used for the clockwork was a telegraphic device, but he used this telegraphic device for non-telegraphic purposes, as he himself confirmed in his 1897 description of the automatic tapper in the Oxford meeting (Lodge 1897, p. 90):

The tapping back was at first performed by hand . . . but automatic tappers were very soon arranged; . . . an electric bell was not found very satisfactory, however, because of the disturbances caused by the little spark at its contact breaker . . . so a clockwork tapper, consisting of a rotating spoke wheel driven by the clockwork of a Morse instrument, and giving to the filings tube or to a coherer a series of jerks occurring at regular intervals . . . was also employed.

The "Morse instrument" was neither a Morse inker nor a substitute for a galvanometer. It was nothing but a clockwork used for tapping. To understand how a clockwork was transformed into a Morse detector, we must examine the impact of Marconi's wireless telegraphy on the British Maxwellians.

Marconi, Preece, the Maxwellians, and "Practice vs. Theory"

Since 1886, Lodge and his Maxwellian friends, Oliver Heaviside (1850–1925) in particular, had been involved in a bitter controversy with William H. Preece (1834–1913), Chief Electrician of the Post Office. At issue were the self-induction of wires and its implications for long-distance telephony and lightning conductors. Heaviside's counterintuitive theoretical claim for the beneficial effect of self-induction for long-distance telephony was severely rebuked by Preece, who based his argument on his practice and experience in the field. In 1887–88 the debate between Preece and Heaviside was intensified by another debate between Silvanus Thomson and Preece concerning telephonic devices. Lodge soon joined the dispute with his innovative designs for lightning conductors, which were severely criticized by Preece. This series of debates evolved into the "practice vs. theory" battle for electrical hegemony.[29] The news that Hertz had discovered Maxwell's electromagnetic waves was known to the British scientific community in 1888. Even though Hertz deprived Lodge (who was attempting to generate and detect electromagnetic waves on wires with discharges from Leyden jars) of credit for the discovery of electromagnetic waves, and even though electromagnetic waves were not directly related to the controversy, Hertz's discovery certainly helped the Maxwellians to defeat Preece. The most important part of Maxwell's theory was proved, and in 1889 Sir William Thomson's warm recognition of Heaviside's mathematical work marked the victory of the theoretical men over practicians.

In February 1896, Marconi came to England with his "secret-box" (figure 2.3). In July, Marconi contacted Preece, who soon became Marconi's first, and most potent, patron. Preece had been interested in induction telegraphy for several years, and thus he may have seen the possibility of commercial wireless telegraphy in Marconi's demonstration. But Preece saw more than commercial possibility; he saw a good means of revenge against the theoretical camp of the Maxwellians. Like Preece, Marconi was "what

Figure 2.3
Marconi in 1869 with his "secret-box" closed (courtesy of Marconi Company Archives, Chelmsford).

Mr. Oliver Heaviside calls a 'practician.'"[30] Marconi knew almost nothing about Maxwell's mathematical theory and perhaps little about Hertz's physical experiments; however, he had developed the Hertzian wave telegraphy, which Lodge had failed to do. To Preece, Marconi's success was a marvelous example of the superiority of practice over theory. The Hertzian wave that had defeated Preece in 1888 now became his weapon.[31]

For a meeting of the British Association for the Advancement of Science that was to be held in Liverpool in September 1896, Preece prepared two papers. In the first (Preece 1896a), which was based on his observations of various submarine cables, he attacked Heaviside's mathematical theory of distortionless cables and advocated his own empirical law. In the second, a discussion of Jagadis Bose's paper, Preece (1896b) stated that "an Italian had come with a box giving a quite new system of space telegraphy"; here, as Lodge later noted (1931, p. 168), Preece was announcing Marconi's success in transmitting a mile and a quarter on Salisbury Plain.[32]

Preece's announcement of Marconi's transmission astonished most Maxwellians, as the following passage in a letter from FitzGerald to Heaviside suggests:

On the last day but one Preece surprised us all by saying that he had taken up an Italian adventurer who had done no more than Lodge & others had done in observing Hertzian radiations at a distance. Many of us were very indignant at this overlooking of British work for an Italian manufacturer. Science "made in Germany" we are accustomed to but "made in Italy" by an unknown firm was too bad.[33]

According to Lodge's later recollection, Lodge did not get up to refute Preece, who was "far more ignorant than he ought to have been of what had been already done," but "retired to [his] laboratory and rigged up an arrangement which [he] showed to Lord Kelvin and a few others, saying 'This is what Preece was talking about.'"[34]

In a public lecture given in Toynbee Hall in December 1896, Preece again publicized Marconi's feat. He also promised that the Post Office would provide unsparing financial support for Marconi's research. This promise upset the Maxwellians, who were then engaged in difficult negotiations with the British government to secure financial support (£35,000) for the establishment of the National Physical Laboratory. Lodge had initiated the latter effort in 1891 at the British Association's annual meeting. When it was revived in 1895 by Douglas Galton, Lodge was appointed secretary of the association's Committee on the Establishment of a National Physical

Laboratory.[35] Besides Lodge (1891b), FitzGerald (1896, p. 383) had also emphasized the importance of science to industrial development. At first the Maxwellians were nervous about Preece, who continually publicized Marconi as the "inventor of wireless telegraphy" and who, in his capacity as Chief Electrician of the Post Office, ignored the importance of scientific research.

At first, their attitude toward Marconi was not very hostile. In March 1897, in a letter to Silvanus Thompson, Oliver Lodge expressed hope that "M[arconi] is improving things all around & going to bring it in commercially." Lodge thought that "there will be many improvements in details wanted before that can be done."[36] But things were moving rapidly. Somebody had coined and publicized the term "Marconi waves," and Marconi approved of it. In an interview with *McClure's Magazine*, Marconi remarked that his wave from the vertical antenna was not same as Hertz's wave. He emphasized that his wave could penetrate almost anything.[37] This strange comment was accompanied by his splendid practical successes. In March 1897, Marconi succeeded in transmitting 4 miles; in May he transmitted 8 miles across the Bristol Channel. Popular reports poured forth, and public interest in wireless telegraphy ran high.

With his "secret-box" and his vertical antenna, Marconi took Hertzian waves out of the scientific laboratories. At first, as is often the case, scientists were not very effective outside their laboratories. Nobody could exactly guess what constituted Marconi's "secret-box." Nobody could explain why the Marconi wave could communicate across buildings and even high hills. Most important, it was not certain why Marconi could send messages several miles when all others had failed.[38] The scientific authorities' ideas about Hertzian waves no longer held. Marconi's practical success and public recognition had established him as the new authority. As an editorial in *The Electrician* remarked, "Professor Marconi," Tesla, and Edison had become authorities on electrical science to the British public, supplanting Lord Kelvin, George Stokes, and Hermann von Helmholtz.[39] Although Marconi's practical wireless telegraphy could also be viewed by the public as a good example of the power of pure science, a subtler battle between theory and practice was underway. As Lodge once complained (1897, p. 91), "the public has been educated by a secret box more than it would have been by many volumes of *Philosophical Transactions* and *Physical Society Proceedings*."

On June 4, 1897, Preece was to lecture on "Signalling through Space without Wires." This was to be the first of the Royal Institution's Friday Lectures to address the subject of wireless telegraphy. Having heard this news, Lodge sent Preece a copy of his Hertz Memorial Lecture of 1894 to "remind him" that Lodge had already worked on the same topic.[40]

In the 1897 lecture, Preece ignored Lodge, compared Marconi to Columbus, and applauded Marconi's feat as "a new system of telegraphy" (1897, p. 476):

> It has been said that Mr. Marconi has done nothing new. He has not discovered any new rays: his transmitter is comparatively old: his receiver is based on Branly's coherer. Columbus did not invent the egg, but he showed how to make it stand on its end, and Marconi has produced from known means a new electric eye more delicate than any known electrical instrument, and a new system of telegraphy that will reach places hitherto inaccessible.

From Lodge's 1894 lecture, Preece quoted Lodge's comment that "half a mile was nearer the limit of sensibility," then proudly declared that "half a mile was the wildest dream." In so doing, Preece successfully derided a theoretician's rash prediction and "scored an effective hit."[41]

The lecture was a blow not only to Lodge but also to most of the other British Maxwellians who had engaged in controversy with Preece several years before. "Preece," FitzGerald indignantly wrote to Lodge, "is distinctly and intentionally scoffing at scientific men and deserves severe rebuke."[42] Lodge was concerned about his credits as a mediator between pure scientific research and commercial wireless telegraphy. In a letter to the *Times*, Lodge explained that the prediction of half a mile was "a scientific one, concerning the small and early apparatus." He emphasized that he had had "the same plan of signalling in 1894." Lodge also emphasized that the same type of coherer that Marconi used had been used by Rayleigh and by Lodge himself.[43] Here Lodge was trying two different but related strategies. The first was to emphasize the essential similarity of his 1894 experiments to Marconi's telegraphy. As Lodge reminded Thompson, "we had the automatic tapping back in '94 at Oxford; . . . we have really had the tapper worked as a relay too & collectors to the coherer; in fact, the whole thing except the best conducting vacuum coherer."[44] Lodge's second strategy was to find the connection between the efforts of British scientists and Marconi's wireless telegraphy. Neither of these two strategies was easy. Lodge's 1894 lectures were not about telegraphy at all, and the connections between the

British scientists and Marconi were too tenuous. Lodge seemed to be excited when he heard that Frederick Trouton, FitzGerald's assistant, had advised Marconi in 1893 or 1894 via one of Marconi's friends. But Trouton's advice proved to be neither of scientific nor of the technical kind.[45] Such efforts, in any case, became meaningless after Marconi's patent was accepted. The impact of Marconi's patent was much more profound than his practical successes.

Marconi's Patent "for Everything"

On June 16, 1897, about 2 weeks after Preece's Royal Institution lecture and 2 weeks before the final acceptance of Marconi's patent, an interesting demonstration was held at a Royal Society soirée. In the entrance hall, Preece and Marconi demonstrated wireless telegraphy in their receptive method of "Signalling through Space without Wires"; on the second floor, Alexander Muirhead demonstrated the same "as practised by Dr. Oliver Lodge in 1894." Muirhead used a Branly tube and a Morse inker; Preece and Marconi used a Morse sounder. The distances between the transmitters and the receivers were about 100 feet. According to *The Electrician*, "Lodge's system worked satisfactorily," and "the marking of the signals on the ribbon were undoubtedly distinct and readable."[46]

From the brief description in *The Electrician* it is evident that Alexander Muirhead had begun to collaborate with Oliver Lodge, thereby initiating competition between Marconi's and Lodge's methods. About a month earlier, Lodge had filed a patent on "Improvements in Syntonized Telegraphy Without Wires." As that title indicates, the principle of syntony (i.e., tuning by varying the inductance of the transmitter and the receiver) was central to the patent. Lodge's patent is now famous as the first patent on syntony, but its provisional specification also claimed that Lodge had improved on Branly's filing-tube coherer and on its use as a detector. Lodge also made a claim on his tapping device. As we have seen, Lodge had employed all these devices in his 1894 lectures. The patent covered the Lodgian system of wireless telegraphy, which had evolved from Lodge's early research on syntony and from his 1894 demonstrations.[47]

Marconi had filed his provisional specification on June 2, 1896, about a year before the filing of Lodge's patent. There was no doubt that Marconi's was the first patent on Hertzian wave telegraphy, but there was much doubt about that patent's efficacy. For Marconi's success to be continued commercially, the patent had to be strong enough to overcome the subsequent litigation. But its provisional specification shows the immature "Marconism" clearly. For instance, the statement "when transmitting through the earth or water I connect one end of the tube or contact to earth and the other to conductors" illustrates Marconi's early conviction that waves from a vertical antenna were different from Hertzian waves.[48] In addition, an automatic tapper of Marconi's own design, operated by the relay current, was described side by side with an independent trembler of Lodge's clockwork type, which undoubtedly weakened Marconi's originality. If Marconi had committed the same errors in the complete specification, he would have invalidated his own patent.

British scientists and engineers generally thought that Marconi could safely patent two things: a tapper activated by the relay current[49] and an antenna (that is, the aerial and the earth connection for the transmitter and the coherer).[50] Except for these two things, the matter was extremely uncertain. His transmitter was of the Righi type, his detector was an improvement on Branly's filing-tube coherer, and his relay and inker were ordinary telegraphic devices. The coherer was the most problematic of these artifacts. Even though the British practice was to award the patent on an invention to the individual who had first applied for it rather than to the person who had first invented the device or published it, it was generally believed that Marconi's claim on the coherer must be a modest one, restricting his claim to the improvement of its sensitivity.

Even expert opinion was uncertain, as is evident from FitzGerald's remarks:

Trouton was sufficiently impressed with [Marconi's secret-box's] value to venture some money in the concern. Since finding out how the thing is really worked he has become much more doubtful as to the validity of the patents and has refused to put any more money into it. It is all a question of patent rights and may depend on such a question as that mercury [in the coherer] is important in order to make the thing work with certainty and that a hammer worked by the relay itself is important and so forth. If these things are of value and patentable, the patents may be of considerable importance. Branly's tube, Righi's emitter &c are all certainly impatentable, but so many things go to make up a workable invention that Marconi's patents may be valuable.[51]

However, FitzGerald's conclusion was optimistic:

As far as I can judge from what I am told it is only details that are patentable and their value is not proved.

The Electrician predicted that Marconi's patent would not be a master patent, insofar as neither the general principles underlying the apparatuses nor the apparatuses themselves were new.[52] And there was another factor contributing to such optimism. Since Marconi was not a man of science, he had probably committed an error in describing the principle of wireless telegraphy (as he had in his provisional specification). If such were case, the patent would be invalidated. At the very least, this might leave room for another patent.

The complete specification for Marconi's patent was filed on March 2, 1897. As Hugh Aitken (1976, p. 204) comments, it was a "different kind of document entirely." Between the provisional and the complete specification, Marconi had secured the crucial assistance of J. Fletcher Moulton, certainly the most famous patent expert in Britain.[53] Moulton's assistance surprised the Maxwellians. Silvanus Thompson, with much surprise, wrote to Lodge on June 30, 1897: "I happen to know that Moulton was called in to advise Marconi on the claim of his final specification of patent, . . . and he advised him to claim *everything*. I understand that as the claim was drawn, they claim, *for telegraphy*, not only coherers, oscillators, & such like details, but even Hertz waves! . . . there is nothing new except the Hertz wave, the oscillator & the coherer, and these are not patented nor patentable."[54] In the interim between the filing (March 2, 1897) and the acceptance (July 2) of his complete specification, Marconi formed a private company to exploit his patent.[55]

As the contents of Marconi's patent were publicized, the "secret-box" was finally opened. (See figure 2.4.) In the patent, Marconi detailed his inventions and attached 19 claims. To everyone's surprise, most of these claims were related to coherers and to various methods of connecting them, such as the ground connection. The claims were not limited to his improvement; they extended to the coherer itself. There were claims on ball transmitters of the Righi type, on a relay, on a hammer-type tapper, and even on improved induction coils and on an antenna (an elevated condenser plate, not a vertical wire).[56] In addition, the awkward expression "through earth and water" was replaced by a more refined expression: "where obstacles, such as many houses or a hill or mountains, intervene between the transmitter and the receiver."[57] FitzGerald noted that Moulton had "drawn [Marconi's] patents too cutely to commit him to any particular theory of what he is doing." Even the critical *Electrician* appraised the specification as "a model of perspicuity."[58]

Figure 2.4
Marconi around 1900 with his "secret-box" open (courtesy of Marconi Company Archives, Chelmsford).

How could Marconi, thought of as a modest and open youth, dare to claim everything about Hertzian waves? How could he claim originality in regard to the Branly tube (which had been used and improved by Lodge) and the Righi-type ball transmitter?[59] Once Marconi's wide-ranging patent was accepted, Lodge had to withdraw his claims on the coherer and on the tapping device in his complete specification (filed in 1898). Only the principle of syntony was left for Lodge to claim.

Lodge must have felt immense frustration, exacerbated by an element of nationalism. Marconi was an Italian. The "ether" had been discovered by great British scientists (Faraday, Kelvin, Maxwell). The Maxwellians, heirs to those scientists, had lost priority for the discovery of electromagnetic waves to a German, Heinrich Hertz. Maxwell's electromagnetic wave was then named the Hertzian wave. Lodge tried to change its name to "Maxwellian wave" at Oxford, but he failed to prevail over the strong objection of another German scientist, Ludwig Boltzmann.[60] Marconi had opened up the possibility of commercial use of the ether, and his comprehensive patent made matters worse. Furthermore, it was clear that wireless telegraphy would have naval and military uses. If his patent went unchallenged, Marconi would monopolize not only Hertzian waves but also important British national interests. Lodge put it this way: "Our old friends the Hertz waves and coherers have entered upon their stage of notoriety, and have become affairs of national and almost international importance." It was thus no accident that, after Marconi's patent, many British scientists and engineers, including J. J. Thomson, George Minchin, Rollo Appleyard, and Campbell Swinton, joined Lodge in deprecating Marconi's originality (Pocock 1988, pp. 103–105; Lodge 1897, p. 91).

As Silvanus Thompson reported in 1899, the patents "were evidently purposely drafted [by Marconi and Moulton] as widely as possible to cover all possible extensions to telegraphy, explosion of mines, and the like, which, indeed, were talked about publicly in connection with Marconi from the first. . . . They are not patents for telegraphy, but for the transmission by Hertz waves of signals or impulses of any kind. . . . In this sense beyond all question Lodge was using Hertz waves for a wireless 'telegraph' in 1894."[61] For Lodge and Thompson, it was Marconi, with his marvelously broad claims, who first violated "the rules of the game." Thus, there was no need for them to follow the rules.

Forging Lodge's Priority

The first evidence used by Lodge's supporters was an article in *The Electrician* titled "Dr. Oliver Lodge's Apparatus for Wireless Telegraphy." The article was intentionally published side by side with Marconi's patent as the "best antidote of Marconism."[62] However, there was in fact no mention of Lodge's telegraphic trial. What the article said was that "Lodge described and exhibited publicly in operation a combination of sending and receiving apparatus constituting a system of telegraphy substantially the same as that now claimed in [Marconi's patent]," and that "Dr. Lodge published enough three years ago to enable the most simple-minded 'practician' to compound a system of practical telegraphy."[63] These are exactly the same as Lodge's two strategies, namely identifying the principles of his experiments in 1894 with those in Marconi's wireless telegraphy and pointing out the possible influence of Lodge on Marconi.

Marconi's position was much strengthened by his overarching first patent, accepted in 1897. In 1898, the "Maxwell-Hertz-Marconi" genealogy in wireless telegraphy was firmly established. Lodge and Thompson tried all possible ways of refuting Marconi. In order to weaken Marconi's patent, they advertised that, owing to the use of wires in the induction coil and other devices, "there is no such thing as wireless telegraphy." They publicized the successes of other scientists, particularly Adolf Slaby (S. P. Thompson 1898; [Lodge] 1898). But, most important for the present discussion, they presented Lodge's 1894 experiments in such a way that those experiments began to be interpreted as telegraphic in nature. This began with the following claim (S. P. Thompson 1898, p. 458):

On several occasions, and notably at Oxford in 1894, [Lodge] showed how such coherers could be used in transmitting telegraphic signals to a distance. He showed that they would work through solid walls. Lodge's great distance at that time had not exceeded some 100 or 150 yards. Communication was thus made between the University Museum and the adjacent building of the Clarendon Laboratory.

In 1900, Lodge admitted that "[Lodge] himself did not pursue the matter into telegraphic application, because he was unaware that there would be any demand for this kind of telegraph."[64] In the third edition of his book *Signalling through Space without Wires*,[65] Lodge's recollection was essentially the same: ". . . so far as the present author was concerned he did not realise that there would be any particular advantage in thus with difficulty telegraphing across space. . . . In this non-perception of the practical uses of wireless telegraphy he undoubtedly erred." (Lodge 1900, p. 45)

Lodge was developing an alternative system of induction or magnetic telegraphy to compete with Marconi's, and in 1898 he allied himself with Preece (who felt betrayed in 1897 when Marconi formed a private company).[66] In 1901, Lodge, then Principal of Birmingham University, launched the Lodge-Muirhead Syndicate.

Lodge's alternative system seemed to have a bright future. The Marconi Company had tried to contract with the Royal Navy, with Lloyd's of London, and with the Post Office, but had encountered a series of obstacles. The Post Office had planned to file a lawsuit against Marconi's company in 1899 and had asked Lodge and Thompson for their expert opinions on Marconi's 1896 patent.[67] (Because Preece was skeptical about the litigation, it was eventually given up.) In 1900, the Royal Navy, which had been suspicious of Marconi's connection with the Italian Navy, prepared more litigation and was given the report of Lodge and Thompson by the Post Office. That litigation too was eventually abandoned after Captain Henry Jackson advised the Admiralty not to pursue it.[68] Lloyd's, instead of contracting with the Marconi Company, had tried to develop its own system. Marconi was stumbling.

The year 1901 was very lucky for Marconi. Lodge abandoned induction telegraphy. The Royal Navy and Lloyd's contracted with the Marconi Company for the use of Marconi's system. In December, Marconi succeeded in the transmitting the signal "SSS" 1800 miles across the Atlantic.[69] After this, Marconi's success was too obvious to be challenged. Lodge and Thompson lost their chance. Moreover, Fleming (Lodge's friend and professor of electrical engineering at University College London) became Scientific Advisor to the Marconi Company in 1899, and FitzGerald died in 1901. The Maxwellian camp was breaking up. The situation became more and more unbearable to Lodge and Thompson.

In April 1902, even before the sensation created by Marconi's first transatlantic success had dissipated, Thompson revived the issue of the invention of wireless telegraphy by attacking Marconi in the *Saturday Review*. He argued that "Signor Marconi [was] not the inventor, but the skilled exploiter, of telegraphy without wires," and that "the original inventor of the wireless telegraphy [was] Professor Oliver Lodge" (S. P. Thompson 1902a, p. 424). As evidence Thompson offered nothing new, only that Lodge had used a ball oscillator, a coherer, a relay, and an automatic tapper in 1894 and had delivered "a signal in the telegraphic instrument." Thompson's motivation for attacking Marconi was also unoriginal; it was his conviction that Marconi, consciously or unconsciously, had devalued scientists' prior credits by violating the rules of the game in 1896–97, and that his successes had been based on this violation.

Lodge wrote to Thompson in appreciation of "the way in which you refer to my claims or rights in the matter." "The opinion of one who is always so well informed on historical subjects," he added, "ought to carry considerable weight."[70]

In his reply, Marconi (1902e) emphasized his priority in patents and the novelty in his antenna and tapper design. Thompson (1902b), in a rejoinder, criticized Marconi again:

Now the matter does not rest on any assertion of mine (for there are scores of persons living who witnessed it) that in 1894 Principal Lodge did publicly transmit signals from one building to another, through several stone walls, without connecting wires, by means of Hertzian waves which were received perfectly clearly upon a telegraphic instrument to which these waves were relayed by means of an automatically tapped "coherer." If that is not wireless telegraphy, then the term has no meaning.

Marconi, who was not skilled in debating and who had not been in Britain in 1894, made no further reply.

In 1906—the year of the second International Congress on Wireless Telegraphy—Thompson reiterated his claim for Lodge's priority, but this time Fleming refuted Thompson's claim:

When it is asserted that Lodge sent "signals" by electric waves in 1894, what it meant is that he caused an oscillatory electric spark made in one room or building to affect a coherer and so move the needle of a galvanometer in an adjacent room, and showed these experiments both at the Royal Institution in June, 1894, and at the British Association, Oxford, in the same year. But there was not a single trace of any suggestion of application to telegraphy in his lecture and in the reprint of it.[71]

In a widely read article ("Wireless Telegraphy") in the 1911 *Encyclopaedia Britannica*, Fleming again refuted the argument that Lodge had demonstrated wireless telegraphy in 1894. However, Thompson's and Lodge's efforts bore some fruit in 1911: Lodge's syntonic patent of 1897 was extended for 7 more years, despite a petition by the Marconi Company. This extension was due in part to Thompson's detailed description of Lodge's wireless telegraphy of 1894 and of Lodge's syntonic patent of 1897 in a memoir (Thompson 1911, pp. 15–16):

In September, 1894, Lodge gave a (third) lecture on Electric Radiation at the Meeting of the British Association at Oxford. This I also recollect very well. . . . Signals were sent to the theatre from the laboratory; and the lecturer pointed out that while a single spark at the sender caused the spot of light of the galvanometer to make a short movement on the scale, a series of two or three sparks such as were made by depressing the key of the sender for a longer time caused the spot of light to make a longer excursion, thus corresponding to the signalling of dots and dashes in ordinary telegraphy. This, so far as I am aware, was the first instance of the public illustration of the wireless transmission of long and short signals.

Encouraged by the extension, Lodge prepared for patent litigation against the Marconi Company. Preece helped settle the case on the condition that the Marconi Company purchase Lodge's syntony patent and appoint Lodge as its scientific advisor. Lodge got that appointment, but his advice was never asked. In the wake of the settlement, the Lodge-Muirhead Syndicate was dissolved (Aitken 1976, pp. 163–168). After that time, the name of Lodge was no longer important in the world of commercial telegraphy.

Thompson died in 1916. Lodge then took up Thompson's role, emphasizing the telegraphic nature of his own 1894 lectures. Lodge later recalled (1925, pp. 37–38):

It was [in 1894] possible to use them [experimental successes] for roughly and imperfectly *transmitting signals in the Morse code,* either by the direct use of a Thomson marine speaking galvanometer or through a relay by operating an ordinary Morse tape instrument. . . . In August 1894, I exhibited *this method of signalling* at the British Association in Oxford.

This claim was modest, but the following year Lodge gave a more detailed and much more audacious description of his Oxford lecture (Lodge 1926, pp. 265–266):

The possibility of actual signalling by this method was insisted on at Oxford. . . . The sending instrument was a Hertz vibrator actuated by an ordinary induction coil set in action by a Morse key. The apparatus was in another room, and was worked by an assistant. The receiving apparatus was a filing tube in a copper hat, in circuit with a battery, actuating either a Morse recorder on a tape, or, for better demonstration, a Kelvin marine galvanometer, as first used for Atlantic telegraphy before the siphon recorder replaced it. The instrument was lent me by Dr. Alexander Muirhead, whose firm habitually constructed a number of cable instruments. . . . When the Morse key at the sending end was held down, the rapid trembler of the coil maintained the wave production, and the deflected spot of light at the receiving end remained in its deflected positions as long as the key was down; but when the key was only momentarily depressed, a short series of waves was emitted, and the spot of light then suffered a momentary deflection. These long and short signals obviously corresponded to the dashes and dots of the Morse code; and thus it was easy to demonstrate the signalling of some letters of the alphabet, so that they could be read by any telegraphist in the audience—some of whom may remember that they did so. Truly it was a very infantile kind of radio-telegraphy, but we found that distance was comparatively immaterial; and at Liverpool, where I was then working, the dots and dashes were received with ease across the quadrangle, or from any reasonable distance.

This became the standard description, and it persisted (even to the present day). The same passage was reproduced in Lodge's 1931 monograph on the history of the British Association for the Advancement of Science, *Advancing Science*, and in his 1932 autobiography, *Past Years*. By the time his autobiography was published, Lodge was over 80 years old, and few living scientists could remember clearly what he had done in 1894. Lodge's argument was picked up by W. H. Eccles, a British radio engineer, and the incorrect diagram of Lodge's 1894 detector depicting a coherer, a telegraphic relay, a hammer-type tapper, and a Morse inker—none of which, except for the coherer, Lodge had actually used—appeared in Eccles's popular book *Wireless* (Eccles 1933, p. 54).[72] Eccles's diagram was noticed by Fleming, who was preparing a lecture on Marconi's life and work. Fleming wrote to Lodge to confirm the diagram, and that was the beginning of the story presented in this chapter.

In this chapter I have dissected the tangled priority dispute between Lodge and Marconi to show how, and in what contexts, Lodge's priority was constructed by his friends and himself over several decades. The crucial issue here—whether Lodge's 1894 Oxford demonstration can be regarded as a wireless telegraphy—hinges, to a degree, on the definition of wireless telegraphy; however, on the basis of various pieces of textual evidence, the assertion that Lodge transmitted and received alphabetic signals does not appear to be true. The claim to the contrary was forged after 1897 as an "antidote to Marconism."

The publicizing of Marconi's powerful patent in 1897 set off a priority battle. However, this alone cannot explain the events that ensued. Lodge's concern regarding priority had been his scientific repute as a mediator between Hertz's experimental physics and commercial telegraphy. Marconi not only patented a device; he also audaciously denied Lodge the very credit he was seeking. Marconi never admitted that Lodge or any other physicist had influenced his work. Another factor was Preece's counterattack. Preece publicly claimed that Marconi's invention was an outcome of the efforts of a practician. This brought all of them (except Marconi, who did not know what Preece was talking about) back to the "practice vs. theory" dispute of several years earlier. This entire sequence of events upset Lodge and other Maxwellians, particularly since they were then trying to show the significance of pure research for technology and industry while seeking governmental support for the establishment of the National Physical Laboratory.

Marconi's "secret-box" threatened the Maxwellians' hegemony in electrical theory and practice. To the Maxwellians, who were always proud of being heirs to James Clerk Maxwell, a genealogical sequence from Maxwell to Hertz to Marconi was unacceptable. Marconi's patent on "everything" monopolized not only the ether but also the use of the coherer, which had been invented and developed by scientists. Nationalism and patriotism, which had been operating rather subtly from the beginning, resurfaced at this point. The British Maxwellians thought that they should not hand over an ether monopoly to Marconi, who they thought had violated the rules of the game. Marconi was not the inventor but only an "exploiter" of wireless telegraphy. It was these complicated contexts that led to the re-characterizing of Lodge's 1894 Oxford demonstration as the first demonstration of wireless telegraphy. His story has been retold many times since then.

3

Grafting Power Technology onto Wireless Telegraphy: Marconi and Fleming on Transatlantic Signaling

If we get across the Atlantic, the main credit will be and must forever be Mr. Marconi's.

—Major Flood-Page to J. A. Fleming, December 1, 1900 (MS Add 122/47, Fleming Collection, University College London)

On December 12 and 13, 1901, in St. John's, Newfoundland, Marconi received messages sent across the Atlantic from Poldhu, England, almost 2000 miles away. He announced his success on December 14. In spite of initial skepticism on the part of some eminent scientists and engineers, Marconi's announcement was welcomed and quickly accepted by the public. However, any electrical engineer who happened to notice the report of Marconi's success in the American journal *Electrical World* (December 21, 1901) would have been very puzzled, because the picture of the "Standard Marconi Apparatus" showed Marconi's table-top telegraphic apparatus, which had not previously achieved transmission over a distance greater than 200 miles.[1] As the editor of the journal commented, it had been expected that the hurdles of long-distance transmission would "be overstepped more gradually."[2] Now, thanks to recent Marconi scholarship, we can understand how deliberately Marconi planned the entire project, and how adventuresome he was in the midst of many difficulties, surmounting various obstacles—natural and technological—with his indefatigable commitment to his project.[3] But, in a sense, we still do not have an answer to the puzzle concerning Marconi's apparatus. Between 1895 and 1901, Marconi had certainly transformed wireless telegraphy from an unstable laboratory experiment into a commercially viable communication technology. Yet his table-top telegraphic apparatus remained essentially unchanged throughout that period. A 10-inch induction coil, a contact breaker, Leyden jars, chemical batteries, and a telegraphic key were invariably used.

In this chapter, I shall highlight the role of John Ambrose Fleming, Marconi's scientific advisor, in helping him achieve rapid transformation of a simple telegraphic device into a powerful system with a 25-kilowatt alternator, 20,000-volt transformers, and high-tension condensers. I aim to correct and reconstruct the history of the first transatlantic experiment on the basis of a detailed analysis of Fleming's unpublished notebooks and other manuscript sources not hitherto cited in the historical literature.[4] After surveying a series of events that led Marconi to his decision to conduct a transatlantic experiment in June 1900, I will analyze Fleming's laboratory experiments in "grafting" power engineering onto wireless telegraphy between July and December 1900—experiments through which Fleming increased his credibility within the Marconi Company. Combining power engineering with wireless telegraphy was neither smooth nor straightforward. Fleming's field experiments between January and September 1901 were essential to the project's success. However, Marconi's intervention in the summer of 1901 was also crucial.

Through detailed examination of Fleming's and Marconi's work, I will also compare two different "styles" of engineering. Fleming, whose educational background was in Cambridge experimental physics, based his approach on scientific engineering—that is, laboratory experiments, precise measurement, and mathematical considerations. Marconi's work was derived from an older style that involved field experiments, handicraft tinkering, and an intuitive understanding of technological effects.[5] Fleming was a trained expert in power engineering; Marconi had created his own expertise in the novel area of antenna technology. Fleming was a university professor and a consulting engineer; Marconi was a "practician" and an engineer-entrepreneur. Their styles could be complementary and creative, but on occasion they were competitive and even antagonistic. They clashed during their experiments at Poldhu, even though they had the same goal. I shall show that the tension between the two men is partially revealed in how credit for the project's success was assigned.

Marconi and Fleming

When Marconi first demonstrated the practicality of his wireless telegraphy before British Post Office officials, in July 1896, the transmission distance was only 300 yards. In his widely publicized demonstration on

Salisbury Plain the following September, he achieved a mile and three-fourths, using parabolic reflectors for both the transmitter and the receiver. Neither the transmitter nor the receiver was grounded. With elevated condenser plates (9 feet high) connected to a transmitter and a grounded receiver, only a third of a mile was achieved. An officer of the Post Office suggested a higher mast (50 feet),[6] and in his second trial on Salisbury Plain, in March 1897, Marconi achieved 4 miles.

During these early demonstrations, Marconi found both the large, elevated condenser plates and the spherical condenser on the top of a mast unnecessary. A long vertical wire was sufficient for long-distance transmission.[7] The height of the antenna and the power of the transmitter were then increased. In May 1897, Marconi experimented at the Bristol Channel, across which William Preece had previously attempted to transmit with an induction device. Marconi used an antenna 110 feet high with a huge cylinder (about $6\frac{1}{2}$ feet high and 3 feet in diameter) on its top. Signals were well received at 3.3 miles and poorly captured at 9 miles (Garratt 1974). In November, Marconi achieved 7.3 miles. Between May and November, several notable events took place: He achieved 12 miles in a demonstration for the Italian Navy. The complete specification of his first patent was accepted and publicized. Having severed his connection to Preece and the Post Office, he formed the Wireless Telegraph and Signal Company to exploit his patent.

In December 1897, the company's first station was built on the grounds of the Needles Hotel at Alum Bay in the Isle of Wight. Soon another was built at Bournemouth, 14 miles away. In April 1898, John Ambrose Fleming went to Bournemouth for a short holiday and happened to visit Marconi's station there. This was Fleming's first meeting with Marconi. Marconi gave Fleming an astonishing demonstration. Even 35 years later, Fleming (1934, p. 116) vividly remembered how surprised he had been to see a Morse printer printing a message ("Compliments to Professor Fleming") that had been transmitted a distance of 14 miles. With Marconi's permission, Fleming inspected Marconi's transmitting apparatus: a 130-foot antenna, a Righi-type ball transmitter, and a 10-inch induction coil. He then examined Marconi's receiving apparatus: a coherer, a Siemens relay, an automatic tapper, and a Morse inker.[8]

Before the advent of Marconi's practical wireless telegraphy in 1896, Fleming had held the Pender Professorship of Electrical Engineering at

University College London for 10 years. He had been only modestly interested in Hertzian waves. "As a pupil of Maxwell," he later recollected, "I
had for long taken an intense interest in his theory of electromagnetic waves
and in the experimental proof given by Hertz in Germany in 1887 of the
existence of these waves . . . ; I had also followed very closely the important
researches of Sir Oliver Lodge . . . and had myself made apparatus for
repeating Hertz's notable experiments." (Fleming 1934, p. 115) Such
research was not, however, a central part of his work, and it was hardly a
substantial basis for his later career as an "ether engineer." In the 1890s,
Fleming was mainly interested in power engineering and low-temperature
physics.[9]

On March 27, 1899, Marconi's first transmission across the English
Channel (between the South Foreland Lighthouse, near Dover, and
Wimereux, near Boulogne, a distance of about 30 miles) caused a nation-
wide sensation in Britain. In one of his first cross-Channel messages,[10]
Marconi apparently invited Fleming to join his camp. Since the founding of
the Wireless Telegraph and Signal Company, in 1897, Marconi had needed
a scientific man of enough authority and influence to cope with the likes of
Oliver Lodge and Silvanus Thompson. In 1898 he had approached Lord
Kelvin, who had paid the company its first shilling for transmitting his wireless messages to George Stokes from the Alum Bay station (an event which
the Marconi Company advertised to show the "commercial" nature of
wireless telegraphy). But after Kelvin, who felt that the "ether" ought to be
shared with the public, asked the company not to raise any more capital, the
company sought other candidates. Marconi approached George FitzGerald,
but he declined (perhaps because of his friendship with Lodge). Although
less famous than Kelvin, Fleming was considered a major authority on both
theory and practice. In addition, he was a close friend of both Lodge and
Thompson, and he was certainly as famous as either of them.[11]

After receiving Marconi's message, Fleming visited Dover to inspect
Marconi's apparatus with his friend Major Flood-Page, a former Managing
Director of the Edison-Swan Company. There he met Marconi and Jameson
Davis, Managing Director of the Marconi Company, and introduced Flood-
Page to Marconi. Soon after this visit, Fleming wrote a long letter to the
Times. After describing Marconi's system very briefly, Fleming, unlike his
scientific colleagues, highlighted Marconi's novel accomplishments in prac-
tical wireless telegraphy. Fleming stated that Marconi had bridged "a vast

gulf [that] separates laboratory experiments, however ingenious, from practical large scale demonstrations," translating "one method of space telegraphy out of the region of uncertain delicate laboratory experiments, and plac[ing] it on the same footing as regards certainty of action and ease of manipulation."[12] Fleming's letter to the *Times* was a blow to Lodge, who had hitherto tried to reduce Marconi's credibility by every possible means. Lodge immediately wrote to Fleming, accusing him of carrying out a public "indictment against men of science, or the Royal Society," to which Fleming replied coolly.[13] Fleming's letter was in fact the first warm recognition of Marconi's work by an important British scientist.

Jameson Davis soon offered Fleming a scientific advisorship in the Marconi Company. Fleming, who had served for about 10 years as a scientific advisor to the Edison-Swan United Electric Light Company and for 5 years as an advisor to the London Electric Supply Corporation, saw this as "a position of trust." He wrote to tell Davis that he wanted to fix his terms "on the assumption that you would possess in return my thought and my inventions or suggestions I may make in your business as your exclusive property."[14] On these terms, Fleming asked £300 per year as a fee and an "engagement for one year certain and renewable year by year and terminable by either side." The company accepted those terms, and Fleming received notification of his appointment as scientific advisor on May 9, 1899. At that time, he regarded this advisorship as just one part of his professional life. He told Davis that he was too busy "to promise you an exclusive attention."[15]

During the first year of his advisorship, Fleming served the company in a variety of ways. First and foremost, he was the bridge between Marconi and the British scientific community. In September 1899, the annual meeting of the British Association for the Advancement of Science was to be held in Dover, and Fleming was chosen to give a lecture in commemoration of the centenary of Volta's discovery of current. During his lecture, Fleming used Marconi's apparatus to send messages from the lecture hall to Marconi's stations at South Foreland and at Goodwin Lightship, which were 12 miles apart and separated by the cliffs of Dover. Messages were also sent across the Channel between the president of the British Association for the Advancement of Science, Michael Foster, and the president of the French Association, Paul Brouardel. The employment of Marconi's apparatus in this demonstration had not only a symbolic value, introducing

Marconi's works to the British scientific community, but also a practical benefit, since Lodge had been advertising that only his magnetic induction telegraphy could transmit across obstacles.[16]

One of the two other research projects on which Fleming worked had to do with the relays used in receivers. Siemens relays, constructed for ordinary telegraphy, were usually used in Marconi's receivers. Fleming sought to devise a relay that would be sensitive enough to work with 0.1 milliampere and that would withstand the vibration caused by the motion of ships. Through the various trials in his laboratory at University College between December 1899 and March 1900, Fleming succeeded in devising a relay that would work with 0.1 milliampere and withstand small vibrations.[17] The second project was a legal effort concerning Marconi's patent. Since 1898, Marconi had worked on preventing interference between different stations, and had succeeded in designing and patenting his syntonic transmitter (patent 7,777, granted in 1900). The essence of the "four-seven" patent lay in connecting a closed condenser-discharge circuit to an open antenna circuit by means of an oscillation transformer called a "jigger." (For details, see chapter 4.) But the patent was bound to encounter troubles, not only because Lodge had previously patented syntonic wireless telegraphy in 1897 but also because Ferdinand Braun of Germany had filed a British patent for a very similar connecting circuit. Therefore, Fleming, with Fletcher Moulton and Major Flood-Page (who joined the board of the Marconi Company as Managing Director in October 1899), carried out a project to strengthen Marconi's complete specification.[18]

These research projects on the relays and on the patent were hardly challenging to a scientist of Fleming's ability. However, Fleming soon found a challenging new project: transatlantic wireless telegraphy.

Building an Engineering Plant in the Laboratory

Even though we do not know exactly when Marconi began to speculate on the "Big Thing" (Marconi's term for transatlantic wireless telegraphy), we know that his understanding that it was technologically feasible was based on his empirical "square law" (which stated that the transmitting distance is proportional to the square of the height of antenna) and on his success, from 1896 on, in transmitting messages between stations whose lines of sight were blocked by hills, cliffs, or buildings, and in transmitting over distances sufficient for the earth's curvature to be significant. Such success, it seemed, could not be explained by the rectilinear transmission of electric waves. In 1896, Marconi thought that his waves actually penetrated obstacles, but he gradually came to regard transmission as occurring along the surface of the earth. In his March 1899 address to the Institution of Electrical Engineers, he asserted cautiously that "electric oscillations are transmitted to the earth by the earth wire . . . and travel in all directions along the surface of the earth till they reach the earth wire of the receiving instrument" (Marconi 1899, p. 280). Marconi's conception of the earth as a waveguide fitted some Maxwellians' conception of the guided propagation of electromagnetic waves.[19] For example, Fleming, who then had no connection with Marconi in March 1899, commented: "I believe also that an important element in his success is due to the earth connection. In his system the spark balls and the coherer are inserted between the earth and the long vertical receiving and transmitting wires. If the earth acts as a perfect conductor the system may then perhaps be regarded as one conductor in which electrical oscillations are set up." (Fleming 1899a) Surface transmission, however, did not automatically guarantee long-distance wireless telegraphy. The whole matter hinged on the power of the radiation and the sensitivity of the detectors.[20]

Around this time, several novel conjectures about the possibility of transatlantic wireless telegraphy emerged. The Chief Electrician of the Marconi Company, James Erskine-Murray, predicted transatlantic wireless telegraphy in an interview with a popular magazine. He told the interviewer that, owing to Marconi's square law and the creeping nature of the Hertzian waves, "if there were another Eiffel Tower in New York, it would be possible to send messages to Paris through the ether and get answers without ocean cables" (Moffett 1899, p. 106). The *Pall Mall Gazette* reported a similar comment from a close friend of Marconi's. Even Silvanus Thompson thought that worldwide communication would be possible in the near future. It is likely that Marconi had long dreamed of this. But this time Marconi immediately denied "any serious attempt being made to establish wireless communication between Ireland and New York." He also expressed the conviction that "real progress could only be made by short stages, and that he was determined to undertake no share in any experiments likely to bring the system into disrepute through complete or even partial failure."[21]

After he became Marconi's scientific advisor, Fleming's attitude toward transatlantic signaling gradually changed. One day in the summer of 1899, Fleming wrote to Marconi: "I have not the slightest doubt I can at once put up two masts 300 feet high and it is only a question of expense getting high enough to signal *to America*."[22] Around this time, Fleming conceptualized the surface transmission of electric waves theoretically by using Joseph Larmor's and J. J. Thomson's electron theory. (For details, see the appendix to this volume.) Marconi also became more explicit about the possibility of transatlantic communication. He made his first visit to the United States between September and November 1899. On his return to England, he began to consider the transatlantic experiment seriously.[23] Soon after he transmitted across the English Channel, in March 1899, Marconi succeeded in transmitting messages between his factory at Chelmsford and Wimereux, a distance of more than 85 miles. In the fall of 1899, however, the British Navy sent and received messages between two ships separated by 85 miles. By the end of 1899, Marconi had not transmitted more than 100 miles.

Objections from the directors of the Marconi Company were numerous. Board members pointed out that increasing power up to several hundred times, which was necessary for transatlantic signaling, was impossible. They argued that such an experiment would interfere with all the other Marconi stations in Britain. Finally, they decided that the experiment would be more harmful than helpful, since the company was in difficult financial straits. No major contract had been signed with the Navy, with the Post Office, or with Lloyd's of London. Although the company's shares had gone from £1 to £6, it was rather like a bubble just before bursting. But Marconi was certain that, with Fleming's help, he could design a power station powerful enough to transmit to the United States. He demonstrated the syntonic experiments to the members of the board in order to relieve their worry about interference. Marconi, in fact, conceived of his experiment as a new corporate strategy. He thought that, if successful, the experiment would eliminate all the company's difficulties once and for all, not merely because the company could monopolize all ship-to-shore communications by showing its capability of long-distance communication in the Atlantic but also because it could compete with long-distance submarine telegraphy. In July 1900 the board assented and the risky experiment was begun.[24]

The whole task was then assigned to Fleming, who thought the first and most important task was to increase the power. Marconi's 10-inch induction coil produced about 100–200 watts and could hardly cover distances greater than 200 miles. (Marconi had achieved 185 miles in early 1900.) An inverse-square relation between the intensity of radiation and the distance showed that transmitting across the Atlantic (2000–4000 miles) would require 100–400 times the energy produced by the 10-inch induction coil. Even before the board agreed to support the experiment, Fleming had thought about the problem of increasing the power. Fleming's theoretical and practical work on a high-tension alternating current system in the 1890s was a good basis for a study of this new problem. In May 1900, when the board of the company was still disagreeing with Marconi on the transatlantic experiment, Fleming wrote to tell Marconi that he had "some ideas for using a transformer instead of an induction coil for long distance work."[25] This letter shows that Fleming, by abandoning the induction coil, had returned to power engineering. In July, Fleming made a list of items he needed for a 25-horsepower station and asked the company for £1000 for their purchase:

A 25 H.P. Oil Engine [18.5 kW]
A 16 Kwatt AC dynamo [21 hp]
A Transformer for raising the voltage up to 20,000 volts
Belts, Switches
A condenser of 1 microfarad capacity
A special AC motor, a rotating commutator[26]

There was no similarity between these items and the apparatus used in ordinary wireless telegraphy. This was the beginning of Fleming's transformation of "laboratory apparatus into engineering plant" (Fleming 1934, p. 118).

In July 1900, a suitable place for the experiment was sought. Marconi selected Niton, on the Isle of Wight, but because a powerful plant might possibly disturb the working of machinery in a lighthouse near it, Fleming urged a different location.[27] Poldhu, on the south coast of Cornwall, was chosen. While Marconi and Page were working out a contract for some land in that region, Fleming bought a second-hand 25-horsepower Mather and Platt alternator and a 32-horsepower Hornsby-Ackroyd engine. He requested two Berry transformers that could increase the voltage to 20,000 volts from the British Electric Transformer Company. He designed a 0.033-microfarad condenser[28] and ordering 36 units from Messrs. Nep Harvey and Peak. The purchases of land and materials were made without the slightest hint of their real intention. Absolute secrecy had to be kept until final success. In August 1900, Richard Vyvyan, a young engineer in the Marconi Company with a background in power engineering, joined the team to assist Fleming and to supervise the construction of a building for the experiment. In December, with the building complete, Fleming sent Vyvyan drawings of the arrangement of machinery based on a series of experiments he had conducted between September and December 1900 in the Pender Laboratory at University College.

Leveraging Laboratory Experiments for Company Politics

Before 1900, Marconi's antenna consisted of a vertical Hertz dipole with the lower portion grounded and the gap placed close to the earth. In 1900, in his "four-seven" patent, Marconi separated the discharge circuit from the antenna circuit and connected them by means of a special oscillation transformer, the "jigger." In Marconi's new system (figure 3.1), the voltage of the battery (a) was increased by the induction coil (C) and charged the condenser (e). The high-frequency oscillation created by the discharge of the condenser across the spark gap passed through the primary coil of the jigger (d), which induced a similar oscillation in the jigger's secondary coil (d') and in the antenna (A–f). To ensure resonance, the syntonic condition ($L_1C_1 = L_2C_2$) had to be satisfied between the inductance and capacitance of the discharge circuit (the loop connecting e, d, and the spark gap) and those of the antenna (E-d'-g-A-f). Marconi also employed the jigger in the receiver circuit to connect its antenna and coherer. When these four separate circuits (the discharge circuit, the antenna circuit in the transmitter, the antenna circuit, and the coherer circuit in the receiver) were all tuned, this new system guaranteed a practical tuning and also a considerable increase in transmitting distance because of the concentration of radiation energy within a narrower range of frequencies. The principle of inductive coupling, as it appeared in Marconi's "four-seven" patent, became an important part of Fleming's long-distance work.

   In Marconi's system, the energy of the induction coil came from dozens of chemical cells. Fleming replaced these low-power components (batteries and induction coils) with alternators and transformers that could supply hundreds of times as much power. This scheme seems simple and straightforward to us, but it did not to the engineers at the beginning of the twentieth

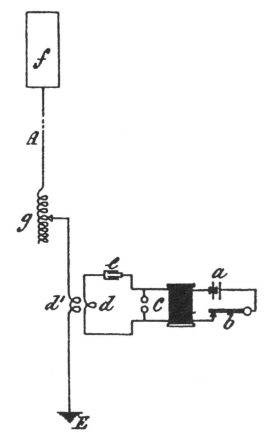

Figure 3.1
Marconi's syntonic transmitter as illustrated in the "four-seven" patent (British patent specification 7,777 (1900)). *e* is the condenser, *C* is the induction coil, *a* is the battery, *b* is the switch, *d* is the primary and *d'* is the secondary of the jigger, *g* is the variable inductance, and *A–f* is the antenna.

century. Sometime in mid 1900, Marconi told Fleming that, by his estimate, a 2-inch spark on a 0.02-microfarad condenser was needed for transatlantic signals. As we will see, Marconi's commitment to 2-inch spark weighed heavily upon Fleming. By a very rough estimation, a 2-inch spark required more than 100,000 volts, since a 1-mm spark was generally regarded as corresponding to 3000–4000 volts.[29] Fleming found from some preliminary experiments that such a high voltage was not obtainable "by the single action of an [AC] transformer." This was the first problem to be solved.

Ultimately he "designed and patented *a method of double transformation*, in which the current from a transformer was employed to charge a condenser and the oscillatory discharge from this sent through another circuit consisting of the primary coil of an oscillation transformer. The secondary of this oscillation transformer was connected to a second pair of spark balls and a second condenser and [another] oscillation transformer, the secondary of the last transformer being in series with the aerial."[30] In this "double transformation" system, a low-frequency (say, 50 Hz) 20-kilowatt alternator charged the first condenser, which had a huge capacitance. From the discharge of this condenser, medium-frequency (say, 10,000 Hz) oscillations were created. Their voltage was increased again by an oscillation transformer to charge the second condenser, which had a small capacitance. Spark discharge from the second condenser finally created high-frequency (say, 10^6 Hz) oscillations, which were radiated into the air through the antenna.[31]

A problem having to do with signaling arose immediately. In his research on the Ferranti transformers in the early 1890s, Fleming had discovered that current and voltage rushed into transformers when they were connected to and disconnected from the main electrical wires, and that this sometimes caused dangerous arcs in the switches. In the case of power engineering, the problem was solved by immersing the switches in oil and by making or breaking the connections slowly. In the case of wireless telegraphy, the keying mechanism created two different problems. First, since the key had to be pressed rapidly and repeatedly to make signals, the keying itself became very dangerous, since the key had to make or break 2000-volt mains. Second, connecting the key at the moment of discharging the main condenser and disconnecting the key at the moment of charging that condenser made the system terribly inefficient. For efficiency, the condenser had to be fully charged, then entirely discharged. Such problems were unprecedented in power engineering (Fleming 1934, p. 110).[32]

Fleming's first "graft" of power engineering onto wireless telegraphy is illustrated in figure 3.2. *A* is a 20-kW alternator. T^1 is a 1:10 step-up transformer, with its primary, *P*, connected to the alternator and its secondary, *Q*, to condenser *C* of huge capacitance. T^2 is a high-frequency oscillation transformer of Fleming's design. Its primary, P^1, connected to *C*, has 100 turns, and its secondary, Q^1, connected to a pair of spark balls D^1 and another condenser C^x with a small capacitance, has 300–400 turns. T^3 is the jigger, a high-frequency oscillation transformer of Marconi's design. Its

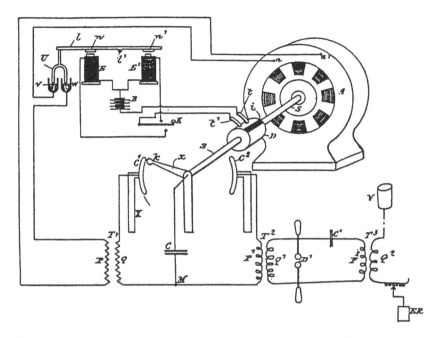

Figure 3.2
Fleming's first "graft" of power engineering onto a high-frequency oscillation circuit.
Source: Fleming, British patent specification 18,865 (1900).

primary, P^2, has one or a few turns; its secondary, Q^2, has several turns. One end of Q^2 is connected to an antenna, V, the other end to earth.[33] The alternator has a frequency of 50 Hz and generates 2000 volts. This 2000 volts is increased to 20,000 volts by T^1. The revolving arm x, when it revolves near C^1, charges the condenser C to 20,000 volts, as the alternator is synchronized to charge C to maximum voltage. The capacitance of C must be very high, since it must take all the energy supplied from the alternator. C discharges when arm x revolves near C^2. Owing to the large capacitance of C (0.5 microfarad) and the large inductance of P^1, a medium-frequency oscillation is created by its discharge. The voltage of the oscillation is raised again by several times by T^2; it then charges C^x. Because of the extremely high voltage, the energy of the oscillation from the discharge of C can then be transferred to the small-capacitance condenser C^x. Owing to C^x's small capacitance (0.02 microfarad) and to the small inductance of P^2, the discharge of C^x can create powerful high-frequency electric oscillations. These oscillations are transferred to the antenna by Marconi's jigger, T^3.

The keying mechanism is a novel feature. K is a key for signaling, and B is a battery for the operation of electromagnets E and E^1. As figure 3.2 shows, when K is pressed down, battery B activates electromagnet E to close mercury switch U; when K is lifted B activates E^1 to break the primary circuit P. Drum D is designed to prevent the keying from interrupting the process of charging and discharging C. In other words, the key is virtually not working when the revolving arm x passes either C^1 or C^2 with the two opposite insulating strips i and i^1 (not shown in the figure) on D. Springs t and t^1 are pressed against D. When t and t^1 are on i or i^1, the battery circuit is disconnected. The lever l is adjusted to remain either up or down. Therefore, whenever the revolving arm x passes C^1 or C^2, the key cannot be connected if it is disconnected, and it cannot be disconnected if it is connected. Pressing or lifting the key does not interrupt the charging or the discharging of the condenser.

Owing to the high voltage, a nearly permanent arc discharge was formed between x and C^1. The arc discharge not only made it difficult to fully charge the condenser C; it also let condenser C discharge before x reached C^2. From September 1900 on, in his laboratory at University College, Fleming experimented to solve this problem. In November he discovered a new method of signaling "in which *no switch* is required *in any of the circuit.*"[34] This method, for which Fleming immediately filed a patent, is illustrated in figure 3.3. As in figure 3.2, A is the alternator, T is a 1:10 step-up transformer, T^1 is Fleming's oscillation transformer, and T^2 is Marconi's

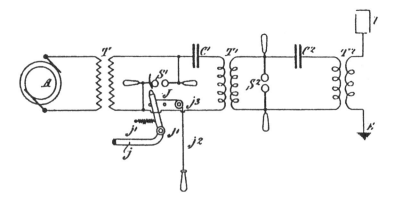

Figure 3.3
Fleming's air blast key. Source: Fleming, British patent specification 20,576 (1900)

jigger. The complex keying mechanism is now absent, the primary circuit of T being always kept closed. The distance between two spark balls, S^1, is kept at 6–7 mm. J is a glass tube jet, pivoted at J^1, through which high-pressure air is blown out. When the alternator is started, it creates a continuous arc discharge across S^1. Pulling down the string j^2 directs J to S^1, and the high-pressure air blast extinguishes the arc discharge. This immediately charges condenser C^1 and then discharges it through sparks across S^1. Owing to the large capacitance of C^1, the oscillation is of medium frequency. It is transformed by the oscillation transformer T^1 to charge the condenser C^2, and the following processes through C^2, S^2, T^2, and I are the same as before. With his air-blast key, Fleming could signal "quite rapidly and well by the mere movement of this jet of air without touching the transformer circuit at all."[35] In a sense, Fleming utilized the arc instead of abolishing it; the arc was transformed into a sort of substitute key that produced signals from the undifferentiated train of waves.

The air-blast key proved to have some defects too. Besides the instability of the air action, it needed extra machinery to produce and maintain high pressure. Moreover, pulling down the string differed from pressing a telegraphic key. Combining the high-pressure air blast with telegraphic keying would require a complex mechanism. These shortcomings made the system, in practice, very inconvenient and unreliable. Fleming therefore redirected his laboratory experiments toward devising a system for use in practice. Two alternatives were open to him. One was to dispense with the air blast but keep the arc mechanism; the other was to dispense with the arc. For the first, Fleming inserted a 10:1 step-down regulating transformer in the primary coil of the main transformer and put two resistances (such as water resistances) in series in the low-voltage circuit of the regulating transformer. He discovered that when the two resistances were connected the discharge across the gap became continuous (i.e., arc discharge), but when these were broken the impedance of the circuit was increased so as to extinguish the arc by reducing the amount of current flowing into the main transformer. The arc's disappearance quickly charged the condenser, which then discharged through sparking, creating powerful oscillations. Connecting and breaking the resistances would be sufficient for a signal.[36] For the second alternative, Fleming put auxiliary condensers of huge capacitance (what he called "arc-stopping condensers") in series with the first discharge condenser. The arc-stopping condenser took some of the applied voltage, so that the arc-stopping condenser "stops almost completely any true arc discharge from the alternating current transformer whilst it permits the active condenser to be charged and discharged with an oscillating discharge."[37]

Fleming discovered these "simple" methods at the Pender Laboratory at the end of November 1900. Though still experimental, they made the whole project much more feasible. In a letter to Marconi, however, Fleming only briefly alluded to his important discovery; he did not give any details.[38] In a letter to Flood-Page, he asked for an improvement in the terms of his contract. Fleming's intention is evident from the two remaining drafts of his letter to Flood-Page. In one draft, after complaining that "in the last six months the work thrown me has grown to such proportion that all my other consulting work (except teaching) is suffering neglect simply because I am overwhelmed all day by things to attend to for wireless," he wrote:

You would hardly believe of the amount of thought this Cornwall exp[erimen]t is requiring. As I said the other day we are engaged in a gigantic exp[erimen]t. It is no routine work like putting up an Electric Light Station. I have to make every step as far as possible sure by laboratory experiment and continual thought by day and night to give it the smallest chance of success.[39]

In another draft, which seems to be the one actually sent to Flood-Page, he was more explicit. After mentioning "a novel dangerous experiment needing every possible precaution," he wrote:

I am willing to do this work on a scale of payment proportional to the responsibility. You are engaged in a gigantic experiment at Cornwall which if successful would revolutionize ocean telegraphy. My view of the case is that if the work asked of me is to continue at the present moment my salary should be raised to £500 per ann[um], as was the case when I was fully occupied for the London Electric Supply Corp.—and in place of a mere three month engagement I ought to have some prospect of reasonable reward (by an increase) if my work and inventions are of material assistance in getting across the Atlantic.[40]

Flood-Page discussed Fleming's proposal with Marconi. Knowing that Fleming's assistance was crucial at the moment, they agreed to present his proposal to the board. On December 1, 1900, Flood-Page notified Fleming that the directors were "prepared to enter into an engagement with you at the rate of £500 a year for three years." However, Flood-Page appended an important qualification: "If we get across the Atlantic, the main credit will be and must forever be Mr. Marconi's." Not foreseeing what it would mean to him, Fleming accepted: "I can confidently leave this [the recognition of his assisting in transatlantic wireless telegraphy] to be considered when the time arrives, assured that I shall meet with generous treatment." Only after this agreement, on December 5, did Fleming (with Marconi's Wireless Telegraph Company) file for patents on the water-resistance key and the arc-stopping condensers. Fleming then sent "a blue print" for the power plant to Vyvyan in Poldhu.[41]

A few days later, Fleming received a "private" letter from Marconi:

Dear Dr. Fleming
I wish you to know of a thing I have had in my mind for some time. You are and have been working so hard helping me in so many ways towards making the long distance trial a success, which as you understand would mean a great deal to me, that I have determined in the event of our being able to signal across the Atlantic, to transfer you 500 (five hundred) shares in Marconi's Wireless Telegraphy Co. Ltd.
I very much hope you will accept my proposal which could be quite independent of anything the Company might think fit to do.
I believe the share will be very valuable if we get across.
In haste
I am yours sincerely
(sigd.) G. Marconi[42]

Marconi's proposal strongly reassured Fleming, for it meant not only considerable compensation but also that Marconi recognized the value of Fleming's work. "When that happy day arrives," he replied to Marconi, "it will be a pleasure to me to accept your gift." Though "we may have great difficulties," he added, "there is certainly in my opinion no impossibility about it and it will revolutionize everything."[43] That "happy day" came just a year later, but during the year Fleming encountered several more problems.

Fleming's Field Experiments at Poldhu

At the end of 1900, the mast made for the Dover experiment (1899) was sent to a small town known as the Lizard, which was 6 miles from Poldhu. George Kemp, who had been working on the syntonic system, was called in to supervise the construction of an experimental station at the Lizard for receiving messages sent from Poldhu as well as the construction of the main antenna of the Poldhu station. Poldhu was far from London and very isolated. In his diary, Kemp wrote the following of his first trip to Poldhu, on January 8, 1901:

I left Chelmsford [where the Marconi Company was located] at 8.5 am for Helston, Cornwall, meeting Mr. Groves [an employee of the Company] at Exeter, and arrived at my destination at 7:30 pm. We went by bus to "The Angel" Hotel where we met

Mr. R. N. Vyvyan who had brought a carriage to take us on to "the Poldhu" Hotel, Mullion, where we dined at 10:30 pm. The ground in the vicinity of the hotel was covered with snow to the depth of 6 inches.[44]

While Kemp was busy building a station at the Lizard, Fleming went to Poldhu for the first time on January 22, 1901. Richard Vyvyan had built the plant there and had arranged the machinery according to Fleming's scheme. Fleming separately tested the alternator, two 20-kW 1:10 step-up transformers, one 20-kW 10:1 step-down regulating transformer, water-tub resistances, condensers, spark balls, and two oscillation transformers installed at the station. The main (first) oscillation transformer was of Fleming's design; it had a primary coil of 20 turns and a secondary coil of 40 turns. The second oscillation transformer, connected to an antenna, was Marconi's jigger. On January 26, Fleming first tested his double transformation system with a water-resistance key (figure 3.4). At first, with the alternator voltage at 1500 volts, he achieved a 6-mm first spark and a 7-mm second spark with the first condenser at 0.3 microfarad and the second at 0.033 microfarad. The power of the primary spark was less than 0.5 kW. He then increased the voltage to 1800 volts and achieved a first spark of 8–9 mm and a second of 11 mm. On January 29, he used two 1:10 step-up transformers to increase the power, with their primaries in series and their secondaries in parallel. Then, by lowering the alternator voltage to 1200–1300 volts, he obtained a 19-mm secondary spark.[45] Marconi also visited Poldhu and witnessed some of Fleming's first experiments. Both Fleming and Marconi were very encouraged to see messages sent from the Isle of Wight coming through at the Lizard, having been transmitted more than 180 miles.

But the 19-mm spark, though much more powerful than those from induction coils, was far from what Marconi was aiming for: a 2-inch spark.[46] Fleming attributed the spark's short length to energy loss from the water resistances. While experimenting, he found a very simple method for using a choking coil in place of water resistance. This is illustrated in figure 3.5, where H^1 and H^2 are choking coils for regulating the voltage with the signaling key K connected in parallel to H^1 and where I^1 and I^2 are E-shaped cores used to adjust the impedances of the choking coils. Calibration proceeds as follows: Remove I^1 entirely from H^1, and, in this position, move I^2 until its impedance is enough to prevent the emf of the secondary coil of T^1 from starting continuous discharge across S^1 but, at the same time, enough to fully charge and discharge the condenser C^1 through sparks. Next, with

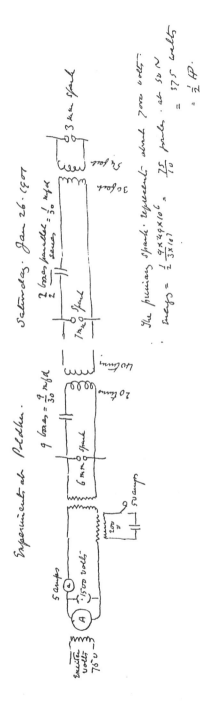

Figure 3.4
Fleming's first test of the double transformation system at Poldhu. Source: Fleming, Notebook: Experiments at UCL and at Poldhu, University College London, MS Add. 122/20.

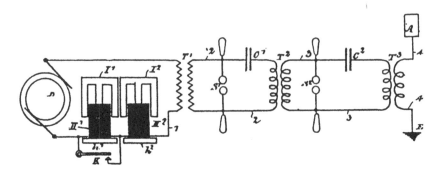

Figure 3.5
Final double transformation system with choking coils. Source: Fleming, British patent specification 3,481 (1901).

I^2 thus placed, put I^1 back into H^1. This increases the impedance of the choking coils enormously, and C^1 consequently ceases to charge at all, terminating any discharge across S^1. Pressing the key K, however, short-circuits H^1, re-creating the spark discharge. Now sparks can be created and destroyed without any arc discharge. In a sense, two of the alternatives in Fleming's scheme of December 1900—utilizing a telegraphic key and abolishing the arc—were now synthesized. As before, owing to the large capacitance of C^1, electric oscillations from the first spark discharge are of a medium frequency, and this voltage is transformed again by means of T^2 to charge C^2 (which has a small capacitance). The final discharge from C^2 through S^2 is of a high frequency and is radiated through T^3 and the antenna A.[47] Fleming specified these choking coils and ordered them from the British Transformer Manufacturing Company immediately. He also increased the number of turns in the secondary windings of the main oscillation transformer from 40 to 320.[48]

In February, Marconi and Vyvyan went to the United States to find a suitable place for a similar power station. Before leaving, Marconi instructed Kemp to order timbers for masts to support the wires that would make up the antenna. Marconi decided on South Wellfleet, Massachusetts, as a site for a receiving station. Vyvyan remained in the United States, and Fleming sent him a detailed memorandum on a power station for the South Wellfleet site.[49] The position corresponding to Vyvyan's in Poldhu was filled by W. S. Entwistle.

Marconi heard that Nikola Tesla, with financial support from J. Pierpont Morgan, was also trying to signal across the Atlantic. "If you can receive there [in the United States]," Fleming reassured the nervous Marconi, "you will establish priority."[50] Marconi hurried. Construction of the antenna began in April. It consisted of a circle of 20 masts, each 200 feet high, with 400 wires connected to them to form an inverted cone. This grandiose design was Marconi's; the construction was supervised by George Kemp. Vyvyan (1933, p. 28) doubted the antenna's mechanical stability, but his objections were ignored.

Fleming began performing his second experiment on April 3, 1901. After the successful test of his choking coil and his new oscillation transformer, Fleming temporarily connected his alternator-transformer system to one 50-foot antenna and began sending messages from Poldhu to the Lizard station, 6 miles away, where they were clearly captured by Kemp.[51] However, once again an arc formed across the first spark gap when the switch was pressed long enough to signal a dash. The voltage across the main transformer increased mysteriously while the key was depressed. Fleming changed the alternator voltage, the alternator frequency, the condenser connections, and the location of the choking coils, but whenever the second spark was longer than 20 millimeters an arc was generated while the key was depressed. On April 18 he traced this to "the large wattless current" (that is, the large phase difference between the primary current and the voltage).[52] The next day, he added inductance to the primary coil of the main oscillation transformer to control this phase difference, and found that "this had a wonderful effect." He got a good 8-mm first spark without any arc, and a 41-mm second spark. He had almost attained his goal. He carefully noted the conditions that brought about such a good result. The number of turns in the primary coil of the main oscillation transformer was 55 or 56. To his surprise, the alternator voltage was only 500 volts. The frequency was 35 Hz. Large inductance, low voltage, and low frequency seemed essential.[53]

After returning to London, Fleming tried to perfect his system. He devised mercury switches for short-circuiting the choke coils, and he had the number of turns in the primary coil of the oscillation transformer increased to 52.[54] On May 22, Fleming made his third visit to Poldhu. Marconi, back from the United States, joined in this experiment. Kemp had erected the first two masts. During this visit, Fleming tested his mercury switches, which proved very satisfactory. He also found that lowering the frequency below 31 Hz caused a very dangerous resonance effect; thus, he changed the armature winding of the alternator, and he increased the number of turns in the primary coil of the oscillation transformer to 85. Employing the same principle he had used in April, Fleming now could obtain a 1-inch second spark. The power of the primary coil was 4.4 kW; that of the secondary coil was 3 kW. On May 28 he noted the following:

In previous expts I have tried using voltage as high as possible [1400–2000 volts]. . . . But in these cases the primary spark soon when 9 mm long was accompanied by great arcing. The spark had a whistling sound and when this was the case we obtained either no secondary spark or else only a brief initial one but could not make dashes. The true secret of success is to use low voltage about 400 and a low frequency say 40 Hz and to have the primary spark gap not more than 6 mm preferably 5.5 mm. Then we can get an inch secondary spark on 1/60 [0.017] mfd.[55]

The situation seemed promising. Once back in London, Fleming instructed Entwistle to change the coil of the main oscillation transformer. On 6 June, he heard that a 26-mm secondary spark had been obtained with a 0.033 microfarad condenser (figure 3.6). The power had been increased by a factor of 6 since April 19. While Fleming was in London working on

Figure 3.6
The transformers and the condensers at work at the Poldhu station around mid 1901. Source: Marconi Company Archives.

a condenser that could withstand high tension, Marconi, at Poldhu, began to experiment on the tuning of the system, using a rudimentary antenna. In mid June, Marconi achieved "good communication" between Poldhu and St. Catherine's station, which were 160 miles apart.[56] Marconi then traveled the 250 miles to Crookhaven, Ireland, to capture the signals from Poldhu. In a letter to Marconi dated 29 June, Kemp wrote: "I am pleased to know you are getting our signals O.K. and hope you will get them stronger when we get the aerial out to its proper place."[57] The equipment and a preliminary antenna were ready for a definitive test.

Marconi Takes Over

On 1 July, Fleming went to Poldhu, where he would spend 10 days working with Marconi. Two 200-foot masts had been erected, and 24 wires, each 200 feet long, had been stretched across them to form a temporary antenna. Fleming used his tuning fork to measure the capacitances of various condensers and aerial wires. On July 4, 1901, they tried "long-distance" transmission between Poldhu and Crookhaven. Since this was the first trial with the antenna designed by Marconi (though only two masts had been built), it was a critical test of the harmony between Fleming's power machinery and Marconi's antenna. All the machinery was adjusted to produce the best signals. Signals were sent from 12:10 to 12:15 P.M., from 12:35 to 12:45 P.M., and at 1:00 P.M.[58]

Was the test successful? Neither Fleming's notebook, nor his *History*, nor any other primary source—even Kemp's detailed diary—seems to give a definite answer. My reading of various sources, however, suggests that Fleming and Marconi did not detect signals (at least in their first series of trials, which began July 4). Four pieces of evidence, though not directly conclusive, support my claim:

• If there had been a successful transmission, surely Fleming, eager to secure his credibility in the Poldhu experiment, would have recorded the success somewhere.

• One passage, in which Fleming deliberately distinguishes Marconi's achievements from his own, suggests that the successful transmission to Crookhaven was accomplished by Marconi alone (Fleming 1906a, p. 451):

In the interests of scientific history, it may be well just to mention briefly the facts and dates connected with the first serious attempt at transatlantic wireless telegraphy. The machinery specified by the author, after consultation with Mr. Marconi, began to be erected at Poldhu in November, 1900, and Mr. Marconi at the same time decided the nature of the aerial that he proposed to employ. . . . In December, 1900, the building work was so far advanced that the writer was able to send down drawings showing the arrangement proposed for the electric plant in the station. This being delivered and erected, experiments were tried by the author at Poldhu in January, 1901. . . . At Easter, 1901, the author paid a second long visit to the Poldhu station, and, by means of a short temporary aerial, conducted experiments between Poldhu and the Lizard, a distance of 6 miles. . . . During the next four months much work was done by Mr. Marconi and the author together, in modifying and perfecting the wave generating arrangements, and numerous telegraphic tests were conducted during the period by Mr. Marconi between Poldhu, in Cornwall, and Crookhaven, in the south of Ireland, and Niton [close to St. Catherine's], in the Isle of Wight.

It is noteworthy that here Fleming attributes success in the transmissions to St. Catherine's and to Crookhaven to Marconi without specifying the exact date.

• Fleming's notebook reports that Fleming and Marconi began checking and modifying the working condition of the Poldhu system after the July 4 experiment.[59] In particular, Marconi changed some essential features of Fleming's design. If the first experiment had been successful, this checking and modifying would not have been necessary.
• In 1903, Marconi recalled: "In October and November, 1901, I succeeded in making 225 miles without the least difficulty between my station at Poldhu and Crookhaven on the west coast of Ireland."[60]

Together, these four pieces of evidence strongly suggest that the July 4 experiment was not successful.

What was wrong with Fleming's powerful system, and how could he find the error? What Fleming had neglected was tuning. He understood well the mathematical principle of tuning ($C_1L_1 = C_2L_2$). However, in practice, the measurement of the capacitance of an aerial was very complicated, and more so the measurement of inductance, which was nearly impossible in many cases. To make matters worse, Fleming's system consisted of two discharge circuits with different characteristics. In figure 3.5, for example, tuning between the primary coil (the discharge circuit of C^2) and the secondary coil (the antenna) of T^3 was possible, because the system had been designed for tuning with Marconi's jigger T^3, the tuning conditions of which Marconi had established well. The real problem lay in tuning the discharge circuits of C^1 and C^2, not only because C^1 differed greatly from C^2 but also because the inductance of T^2 (Fleming's oscillation transformer) could not be estimated. Of course, there was no wave meter at that time.

Fleming might have considered the tuning of the two circuits across T^2 unimportant, since the oscillatory discharge from C^1 was of only medium frequency.

On July 8, Fleming used his "tunmeter" (in fact a hot-wire ammeter) to determine the tuning of various circuits by measuring maximum current, and concluded that the aerial was not in tune with the discharge circuit. But Marconi had a different opinion and proposed a radical solution. He replaced Fleming's main oscillation transformer with his jigger in order to bring the two discharge circuits into tune with each other (figure 3.7). On July 10, with Marconi's jigger in the circuit, they "carried some good experiments" by varying the capacitance.[61]

After the July experiment, Marconi actively intervened in the Poldhu experiment. Fleming went back to London, and Marconi went to Crookhaven again, having had Kemp manipulate the power machinery at Poldhu. At the end of July, Marconi himself increased the length of the spark by changing capacitance and voltage. Then, Marconi decided to rearrange the machinery at Poldhu. He built a new house near the old one and placed the condenser and jigger in the new house, separating them from machinery. This act symbolically represents Marconi's monopoly of power over the project. Marconi, and only Marconi, could manipulate his jiggers for tuning yet. Marconi had access to Fleming's power machinery. Fleming's role was now reduced to rebuilding broken condensers and rewinding the armature of an alternator for Marconi. The machinery was rearranged between August 2 and August 20, 1901. On August 23, Fleming went back to Poldhu and tested this new arrangement. On September 2, he finished his testing, "left it all ready for Mr. Marconi," and left for Paris on a vacation.[62] The antenna was almost complete.

On September 17, the 20 masts and 200 wires at Poldhu collapsed in a bad storm. Shortly thereafter, all the antennas at Wellfleet suffered the same misfortune. Marconi decided to quickly build two new masts at Poldhu to form a fan-shaped antenna with only 54 wires. His plan for reciprocal communication between Poldhu and Wellfleet was altered to unidirectional sending at Poldhu and receiving at any place in the United States. Marconi's quickly fabricated antenna was vastly inferior to his first design, but nevertheless he re-tuned the system for experiments. In October and November, signals were clearly received at Crookhaven. The receiver that Marconi used in these experiments, a new mercury coherer with a telephone, had been

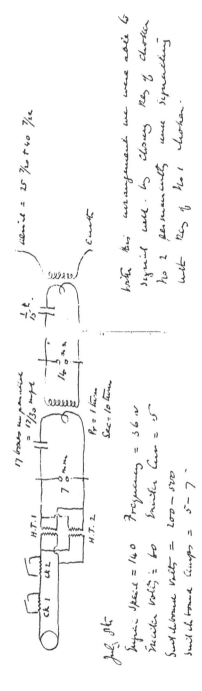

Figure 3.7
The Poldhu system after July 8, 1901. Note that Marconi's jigger has replaced Fleming's oscillation transformer. Compare this with figure 3.4. Source: Fleming, Notebook: Experiments at UCL and at Poldhu, University College London, MS Add. 122/20.

given to Marconi in August 1901 by his friend Luigi Solari, an officer in the Italian Navy. It turned out to be very efficient.[63]

The Poldhu-Crookhaven experiment was so successful that Marconi decided to perform a transatlantic experiment. He chose St. John's as a reception site because of its proximity to England. With his assistants George Kemp and P. W. Paget, and with two balloons, six kites, coherers, and other receiving devices (including the "Italian Navy coherer"), he left England on November 27, 1901. Before leaving, he wrote the following in a letter to Entwistle:

When I wish you to start sending across, I shall telegraph the date to the office in London, this date will be immediately wired on to you and from this advice you are to start sending the dot programme from 3 p.m. to 6 p.m. Greenwich time [which corresponded to 11:30 am to 2:30 pm in St. John's time], and are to continue sending same programme during the same hours every day (Sundays excepted) until you are instructed to stop. . . .[64]

In Newfoundland, Marconi pretended to be researching the influence of coastal rocks on the Hertzian waves.[65] The rest of the story, which has been retold in many works on Marconi, is well known. On November 9, Marconi cabled instructions to the London office to transmit "SSS"[66] between 3 and 6 P.M. each day from December 11 on; a telegram "Begin Wednesday 12th" was sent from the London office to Entwistle at Poldhu. Marconi's first trial with balloons failed on the 11th (a Wednesday) when a strong wind carried the balloon away. Despite the bad weather, Marconi and Kemp, using a kite aerial, a mercury coherer, and a telephone, received the signals in St. John's on December 12 (at 12:30, 1:10, and 2:20 P.M.) and December 13 (at 1:38 P.M.). Since further experiments were ruled out by the weather, by the lack of proper equipment, and by the reluctance of the American telegraph company to help, Marconi arranged a press announcement for December 14.[67]

That Marconi heard the "SSS" signal was attested only by his and Kemp's testimony. Skeptics argued, not unreasonably, that Marconi was deceived by stray atmospheric noise or by signals from nearby ships. Famous scientists and engineers, including Thomas Edison, Oliver Lodge, William Preece, and Edouard Branly, strongly doubted Marconi's claim. John Ambrose Fleming, Michael Pupin (scientific advisor to the Marconi Company of America), and Elihu Thomson, however, supported Marconi. Lay opinion, and that of many newspapers and public magazines, was in favor of the young hero. For reasons unknown, Edison later quickly changed his mind. His support was extremely helpful. On December 21, 1901, the *Times*, which had been skeptical of Marconi's announcement, published a warm editorial comment: "It would probably be difficult to exaggerate the good effect of wireless telegraphy, if, as Mr. Marconi and Mr. Edison evidently believe, and the Anglo-American Company evidently fear, it can at no distant time be developed into a commercial success."

In January 1902, the American Institute of Electrical Engineers announced a special dinner in Marconi's honor. This marked the beginning of a shift in his favor.[68]

"Main Credit Must Be Forever Marconi's"

What did Fleming do at this crucial moment? According to the engineer-historian George Blake (1928, p. 101), "Fleming had transmitted the letter 'S' from Poldhu, in Cornwall, to St. John's, Newfoundland." This incorrect claim was magnified by the writer David Woodbury (1931, pp. 146–147), who filled in missing details with his imagination:

With what pounding of the heart must he [Marconi] have watched his little coherer attached to the lower end of the wire! Like Franklin, he was determined to steal from the clouds another great secret of electricity. All at once the filings in the coherer stiffened. Something was coming—yes, it was Fleming! . . . Faintly, almost too faintly to be understood, came the letter "S" of the Morse signal . . . Marconi had won. The pressure of Fleming's hand upon a key had been heard two thousand miles away.[69]

Fleming had not even been at Poldhu. He had stayed in London, fulfilling his professorial duties and preparing his Christmas lecture for the Royal Institution. He had not been informed when the signaling began. He had been completely excluded from the experiment since September 1901. He was "left in ignorance of this success until [he] opened the *Daily Mail* newspaper in London on the morning of Monday, December 16, 1901" (Fleming 1934, p. 124). Some days later, he asked an employee of the Marconi Company about the arrangements for the machinery actually used at Poldhu on December 12; he recognized that these "were substantially as I left them in September, 1901" (figure 3.8).[70]

As time passed, Fleming felt more and more frustrated. The *New York Herald* quoted Marconi as follows: "Before leaving England I arranged for our long-distance station." In January 1902, when American electrical

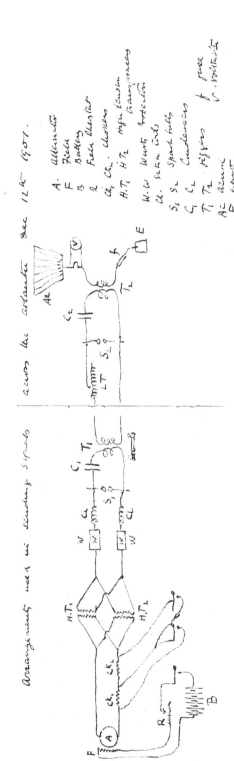

Figure 3.8
Fleming's later drawing of the arrangement of machinery at Poldhu that was actually used for Marconi's first transatlantic transmission on December 12, 1901. Source: Fleming, Notebook: Experiments at UCL and at Poldhu, University College London, MS Add. 122/20.

engineers held a congratulatory dinner for him, Marconi expressed his debt to James Clerk Maxwell, Lord Kelvin, Joseph Henry, Heinrich Hertz, and Alexander Graham Bell (for the telephone detector). Fleming was spoken of only as one of several company employees who had assisted Marconi. In subsequent articles, reports, and speeches, Fleming's name was never mentioned. Even after his return to England, Marconi attributed everything to his own decisions and achievements.[71] As Flood-Page had told Fleming a year earlier, the main credit was "forever Marconi's."

Fleming's involvement in this ground-breaking experiment did receive some recognition, however. *The Electrician* of March 7, 1902, noted Fleming's assistance "in advising on the selection and arrangement" of the Poldhu station. It stated that Fleming had been responsible for "designing the details of the motive power and electric machinery for generating and controlling the powerful electrical oscillations used to produce the electric waves," and that "he was down at Poldhu for many weeks during last year superintending the testing of the plant, and experimenting, alone and in company with Mr. Marconi, on its capabilities."[72] This portrayal of events provided an opportunity for Marconi's enemies. In April 1902, Silvanus Thompson, still friendly with Fleming, attacked Marconi by pointing out that the Poldhu station had been designed by Fleming, not by Marconi, and that the detector Marconi had used in Newfoundland was not of his own design (S. P. Thompson 1902a, p. 425).[73] In discussing their response to Thompson, Fleming complained to Marconi about the failure to share credit. Marconi replied:

It is needless for me to say that I had and have every intention to make reference to your work and assistance in connection with the Poldhu plant . . . had I not refrained from doing so at your own request. . . . At the present juncture it would greatly facilitate my statement on your assistance in the transatlantic experiments if you could also see your way as a scientific man to protect me in the same manner as you have done in the past from the violent and very often unfair attacks to which I have been lately so much exposed.[74]

Marconi had planned to give a Friday Lecture on his recent work on wireless telegraphy at the Royal Institution on 13 June. Since Friday Lectures were always influential, it would be a good forum in which to credit Fleming. Fleming specified: "You may be quite sure that a generous mention at the Royal Institution of my work for you will not only please me and my friend very much but will call for an equally public acknowledgement on my part of the fact that the achievement is essentially due to your inventions."[75] "My friend" here cannot be anyone other than Silvanus Thompson. Marconi modestly kept his promise during the Friday Lecture, mentioning that "the general engineering arrangements of the electric power station erected at Poldhu for the execution of these plans and for creating the electric waves of the frequency which I desired to use were made by Dr. J. A. Fleming" (Marconi 1902d).

Consoled by Marconi's recognition, Fleming tried to take a part in improving the Poldhu station for more stable transatlantic communication. Marconi, with help from Vyvyan and Entwistle, had already changed Fleming's double-transformation system to a single-transformation one, and had put up larger and more stable antennas. In July, Fleming went to Poldhu to supervise a long-distance experiment between Poldhu and the *Carlo Alberto*, an Italian Navy ship that had been placed at Marconi's disposal. The *Carlo Alberto* was sailing from Poldhu to St. Petersburg, Russia. Marconi, aboard the *Carlo Alberto*, was worried about Fleming's modification of the Poldhu system. To Marconi's chagrin, Fleming had installed a new rotating disk discharger of his own design for this experiment and had changed the single-transformation system to a double-transformation system. The signals sent from Poldhu were received at Kiel, Germany, more than 600 miles from Poldhu. However, at a crucial moment when the Russian emperor was visiting Marconi's ship, Marconi was unable to receive signals, and he had to send signals from another room of the ship.[76]

Marconi, already vexed by Fleming's game playing in regard to Thompson's attack, wrote to H. Cuthbert Hall, then Managing Director of the company, that "the revolving disc dischargers designed by Dr. Fleming for this station have proved in practical working to be unsatisfactory," and that "Dr. Fleming seems to introduce so many complications which in practice prove useless, that I think it will be well that the details with regard to the changes in the plant here [at Poldhu] . . . should be discussed and settled between you and Mr. Entwistle, as I am afraid that no useful purpose would be served by referring them to Dr. Fleming."[77] Marconi ordered Hall to change some of the equipment at the Poldhu station, and the changes included installing rotating dischargers of Marconi's own design. Fleming resisted this change, which disturbed Marconi again. Marconi complained to Hall:

[Dr. Fleming] informed Mr. Entwistle that the designs ought, as a matter of courtesy, to be referred to himself for approval before being sent to the office. This attitude on his part opens up again the wider question of his general position in the Company and I am desirous that this should be clearly defined to him without further delay. It should be explained to him that his function as Consulting Engineer is simply to advise upon points which may be expressly referred to him and in no way places upon the Company any obligation to seek his advice upon any matters in which it is deemed unnecessary. In this particular case I can see no reason for consulting him whether based on courtesy or any other consideration. He was asked in the first instance to prepare a design and, this having proved unsatisfactory, I dealt with the matter myself in conjunction with Mr. Entwistle. I do not wish to inflict any unnecessary wound on Dr. Fleming's susceptibilities, but, unless you are able to put the matter before him effectively in a right light, I shall feel bound to make a formal communication to the Board with reference to his general position.[78]

Marconi's neglect of his earlier promise to give Fleming 500 shares in his company may have been due entirely to his busy schedule, but it deeply hurt Fleming, who never forgot this important breach of trust. It seems that Fleming reminded Marconi of the promise in February 1903, and Marconi delivered the shares.[79] Fleming seems to have remained bitter that credit for the first transatlantic communication had gone to Marconi, though he did not complain about that until after Marconi's death in 1937. Marconi, Fleming told Lodge, "was always determined to claim everything for himself. His conduct to me about the first transatlantic transmission was very ungenerous. I had planned the power plant for him and the first sending was carried out with the arrangement of circuits described in my British patent No. 3481 of 1901. But he took care never to mention my work in connection with it."[80] As was noted in chapter 2 above, Fleming even declared that Marconi could not be said to have invented wireless telegraphy. That is now more understandable.

Marconi's and Fleming's Differing Research Styles

Fleming's frustration and Marconi's vexation had deep roots in their differing styles of research. Fleming had been trained under James Clerk Maxwell at the Cavendish Laboratory in Cambridge, where he had learned the importance of combining precise measurement and mathematical considerations in physical research.[81] During the years 1882–1899, Fleming applied his scientific knowledge and methodology to heavy-current engineering, establishing a solid reputation among both electrical engineers and physicists. After entering the field of wireless telegraphy, he applied his methodology and knowledge of scientific engineering to technology. In the

Poldhu experiment, as we have seen, Fleming began by setting up the power machinery, then measured the electrical quantities he was accustomed to measuring (such as the primary and secondary currents) as well as the voltages, capacitances, resistances, and alternator frequencies. With numerical data and theoretical formulas in hand, he progressively increased the length of the secondary spark to 2 inches.

Marconi had little training in physics. In Britain, he was thought of as a practician. From the beginning of his career, Marconi taught himself by means of numerous trials. In designing wireless circuits and other devices, Marconi employed a rigorous personal logic and a kind of talent that neither scientists nor science-oriented engineers seemed to possess. As Fleming once stated, Marconi "did not arrive at any of his results by mathematical prediction. In fact I think his mathematical knowledge was not very great. . . . In addition to this power of intuitive anticipation he possessed enormous perseverance and power of continuous work."[82]

Marconi's starting point for the Poldhu experiment was antenna design, a field in which he was certainly competent. Initially, he left the power machinery to Fleming. Marconi, it seems, gradually came to understand that Fleming's "very dangerous" power machinery was really nothing but a sort of substitute for an induction coil and chemical batteries, and that the principles of wireless telegraphy were not involved at all. These principles remained, Marconi thought, his own possession.

The "July failure" happened at a crucial moment. If the principles were the same, Marconi might have reasoned, why not try his jiggers in the new power system? As we have seen, the jigger radically changed the direction of the Poldhu experiment, because Marconi, not Fleming, was able to manipulate the special relations between the forms of coils and the numbers of primary and secondary windings.[83] After the introduction of the jigger into the Poldhu system, Fleming's role was destined to be minimized.

Various old and new elements of engineering are represented in the Fleming-Marconi story. Fleming represented an established branch of electrical engineering: power engineering. The jigger and the antenna, Marconi's contributions, belonged to the emerging field of wireless telegraphy. Fleming's scientific engineering was a relatively new approach to practical engineering, having originated with William Rankine and William Thomson (later Lord Kelvin) in the middle of the nineteenth century and flourished in the 1880s and the 1890s in the work of engineers such as

Alexander Kennedy, Fleming, and John Hopkinson (Buchanan 1985). In comparison, Marconi's method—associated with famous mechanical, civil, and electrical engineers of the eighteenth and nineteenth centuries, including James Watt, James Brindley, M. I. Brunel, R. E. B. Crompton, and Sebastian Ziani de Ferranti—was rather old and traditional.[84] And, although Fleming's position as a university professor in engineering was relatively new, the position of a consultant engineer to private firms was quite a traditional occupation in the British engineering history. In contrast, although Marconi's position as an independent inventor was well established, his position as an entrepreneur-engineer in an international corporation was unprecedented in British engineering history.

These different styles of engineering clashed in the case of the Poldhu experiment, and the differences were especially evident in the tension over credit. Fleming had good reason to regard the painstaking experiments on power machinery he conducted from September 1900 to September 1901, which made the double-transmission system stable, as essential to Marconi's success. Fleming approached the Poldhu experiment as an experimenter. To him, the antenna, the tuning, the detectors, and other devices were auxiliary to the power component of the transmitter. To Marconi, power machinery was just one component of his system for long-distance wireless telegraphy; other components were equally important. As he told Fleming, Marconi took "credit for the essential necessary arrangements which [he] devised for use in conjunction with large power plants, the tuning of the different circuits in the transmitters and receivers, the determinations of the shape and size of the aerials and the special forms of oscillation transformers which have proved so successful in practice."[85]

Fleming was a hired consulting engineer beset by many constraints. For example, he had to satisfy Marconi's request for a 2-inch spark. This required more than 100,000 volts, and such a high voltage required a double-transformation system. For the double transformation, Fleming employed another oscillation transformer, of his own design, whose inductances could not be estimated. Owing to this defect, syntony between the first discharge circuit and the second was nearly impossible. (In July 1901, as we have seen, Marconi substituted the jigger for Fleming's oscillation transformer.) Furthermore, the system had difficulty producing dashes without generating an arc across the spark gap. Owing to this flaw, only dots were transmitted in Marconi's final trial. This was far from a stable system for use in commercial telegraphy.

Fleming's double-transmission system was soon replaced by a single-condenser system similar to Marconi's "four-seven" circuit (Marconi 1908, p. 117; Vyvyan 1933, p. 35). Alternators much more powerful than the 25-kilowatt unit at Poldhu, including a 75-kW one at Grace Bay (1902), a 150-kW one at Grace Bay (1904), and a 300-kW one at Clifden (1906), were later employed. Fleming's glass condensers were replaced by huge air condensers (Clifden in 1906). Automatic signaling machinery was also employed in the new stations in the United States and Ireland. The new "single-transmission" system worked well. Fleming's original double-transmission system originated not merely from technical necessities, but also from Fleming's commitment to achieve, with relatively low-power machinery, Marconi's goal of a 2-inch spark. Because the double-transmission system at Poldhu was troublesome and replaceable, Marconi regard Fleming's contribution as less than essential.

The difference between Fleming's and Marconi's professional styles deepened their mutual misunderstanding. Fleming's incentive to do consulting work had two elements. First, he was able to procure not only apparatus, machinery, and some funds but also the "field" for teaching and research from the electric supply companies. Second, he was able to increase his professional credibility in the community of electrical engineers and physicists by combining field experience with scientific knowledge. His earlier scientific advisorships had served him well in this regard. His consulting work for the London Electric Supply Corporation and for other companies enabled him to solve the mysteries of the Ferranti effect and of the efficiency of transformers.[86] His advisorship to the Edison-Swan Company provided him with an opportunity to investigate the Edison effect in incandescent lamps—a research that made him a fellow of the Royal Society in 1892. He expected his involvement in the Poldhu experiment to bring about similar results.

Marconi was both an independent engineer and an organizer. He employed every useful resource around him. Fleming contributed to the transmitting machinery; Vyvyan and W. H. Eccles contributed much to the improvement in the design of Marconi's jigger; Solari provided the most sensitive detector; Marconi himself designed the antenna and the tuning system; Kemp built the antenna; Entwistle operated the transmitting machinery; Marconi and

Kemp overcame the weather in Newfoundland. The experiment was one of collaboration. Marconi performed the role of organizer, and he strongly felt that it was to the organizer that the highest credit should be assigned. Fleming, though very helpful, was only his advisor.

Marconi was also a businessman. Marconi decided to begin the transatlantic experiment at a time when his company had come to a commercial dead end. The experiment was conceived of as a stepping stone to his monopoly of Atlantic ship-to-shore communications, and ultimately to a larger scheme of world communications. It was conceived of to provide a breakthrough for the negotiations with the Navy, with Lloyd's of London, and with the Post Office. At the same time, it was a weapon to quiet his enemies and an opportunity to increase the capital of the company. The company's fortune relied heavily on Marconi's authority and achievement. His desire to monopolize credit for the first transatlantic experiment seems to have originated as much from his business plan as from his personal acquisitiveness. Fleming, as a consulting engineer, never could have understood this.

4

Tuning, Jamming, and the Maskelyne Affair

There was a young fellow of Italy
Who diddled the public quite prettily
—Nevil Maskelyne, June 4, 1903 (Blok 1954)

Witnessing and reporting of successful experiments were strategic for Marconi. They advertised the practical nature of his system. Marconi's Salisbury Plain demonstrations in 1896 and 1897, his demonstration across the Bristol Channel in 1897, communication in 1898 between the Royal Yacht *Osborne* and Osborne House in the Isle of Wight, and the transmission of messages across the English Channel in 1899 and across the Atlantic in 1901 widely publicized his wireless system. The event that the famous and respected Lord Kelvin paid a shilling to send messages to William Preece and George Stokes was advertised as showing the commercial feasibility of Marconi's wireless system. This deeply impressed the public; however, it brought Marconi and his company into direct conflict with the British Post Office, which monopolized all forms of commercial communications in the Empire.

When syntony (i.e., tuning) became a central issue in radiotelegraphy, Marconi devised a new system for sending and receiving messages. This "four-seven" system, as we saw in the previous chapter, was employed in Marconi's transatlantic experiment. Marconi also held a series of demonstrations of the system's effectiveness. These were witnessed and reported by Marconi's scientific advisor, John Ambrose Fleming. Fleming was able to act as a trustworthy witness because of his high credibility in the British electrical engineering and physics communities—a credibility he had built up over 20 years by serving as a mediator between alternating-current power engineering and physics.

Fleming, as a supportive witness, was troublesome to Marconi's adversaries. In June 1903, Nevil Maskelyne, one of Marconi's opponents, interfered with Fleming's public demonstration of Marconi's syntonic system at the Royal Institution by sending derogatory messages from his own simple transmitter. This incident, which became known as the Maskelyne affair, severely damaged both Marconi's and Fleming's credibility. Indeed, Fleming was dismissed from his advisorship to Marconi soon after the affair.

The Maskelyne affair has never been fully examined in all the relevant contexts (the scientific and technological context, in which early syntonic devices were understood and developed; the corporate context, in which the competition for monopolizing the market for wireless telegraphy became intense; the authorial context, in which questions as to who should hold authority in the rapidly developing field of wireless telegraphy became prominent). This chapter aims to give a detailed examination of the Maskelyne affair within these contexts. It will uncover several issues worth exploring, such as the struggle between Marconi and his opponents, the efficacy of early syntonic devices, Fleming's role as a public witness to Marconi's private experiments, and the nature of Marconi's "shows." In addition, the affair provides a rare case study of the manner in which the credibility of engineers was created, consumed, and suddenly destroyed.

Syntony and the "Four-Seven" Patent

As we saw above, Marconi's early system deployed vertical antennas, with one pole grounded. Marconi achieved a range of several miles with this system, but it was difficult to tune. In practice, bad syntony meant that anybody could easily capture messages with a simple detector. In theory, it meant that the waves generated by the transmitter did not have a narrowly defined frequency range. According to Hertz, the frequency (f) of the wave generated from a condenser-discharge circuit was (very nearly, at least) determined only by its inductance (L) and its capacitance (C), according to the equation

$$f = \frac{1}{2\pi\sqrt{LC}}.$$

The wave generated by Marconi's vertical antenna transmitter should thus have had a well-defined frequency determined by its own inductance and capacitance, but that did not seem to be the case. A consensus emerged among physicists and engineers that the major reason for this was that the waves generated by Hertz's and Marconi's devices were highly damped. Although at first opinions concerning the reason for the damping varied, it was eventually decided that a highly damped wave could be thought of, physically and mathematically, as a superposition of many different waves, each with its own frequency. Indeed, according to Fourier analysis (which was granted direct physical significance here), the more rapidly a wave is damped the broader is its frequency range. This "multiple resonance" was regarded as one of the most difficult problems in Hertzian physics.[1] The more homogeneous (or "continuous," as it was described by engineers of the day) the wave, the narrower its frequency range. Therefore, in order to secure good syntony one had to produce (or come close to producing) homogeneous or continuous waves.

Though it was not possible to produce continuous waves by means of a spark transmitter, there was a way to lessen damping. Since the early 1890s, physicists had known that a transmitter in the shape of a loop was a more persistent vibrator than was a vertical antenna. In early 1897, while communicating with Silvanus Thompson on Marconi's wireless telegraphy, Oliver Lodge was reminded of his 1889 work on producing syntony between Leyden jars, and of its practical implications for wireless telegraphy.[2] Those experiments had employed a species of closed circuit that Lodge now adapted to the demands of wireless signaling. He applied for a patent on such a circuit in May 1897.[3] In the application, Lodge described two principles. First, the damping of radiated waves was reduced by means of a closed-circuit transmitter, whose sparking surfaces were partially enclosed in a metallic box. Second, to receive a wave with a specific frequency, Lodge inserted a variable inductance into the receiver in series with the capacitance of the receiver's antenna (figure 4.1).

Marconi paid little attention to syntony at first. When he filed a provisional specification (June 1896) and a complete specification (March 1897) for his first wireless patent (12,039), he did not mention syntony at all, though there is evidence that he knew about it in 1896.[4] Nor, at first, did syntonic issues trouble Marconi much. Owing to the nature of his transmitter, the emitted wave was highly damped, like Hertz's. As a result, the wave acted as if it were a mixture of waves having a broad range of frequencies, and this made Marconi's untuned receiving antenna respond extremely well.

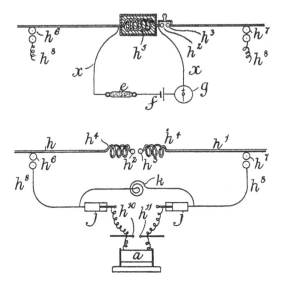

Figure 4.1
Lodge's syntonic transmitter and receiver employing variable inductance. Source:
O. Lodge, "Improvement in Syntonized Telegraphy without Line Wires," British
patent 11,575, filed in 1897.

Before long, however, the market for wireless telegraphy forced Marconi
to face up to syntonic issues. Initially, the demand for wireless telegraphy
came from the military (the Army, the Navy, and the War Office), mainly
because there were no wires that the enemy could cut down. Marconi had
been well aware of this, and of the importance of secrecy. The only way to
secure secrecy seemed to be syntony. In January 1898, Lodge's syntonic
principle and his closed transmitter with variable inductance in the receiver
were made public in a paper Lodge read before the Physical Society.[5] Talk
about syntony became common. Adolf Slaby (1898) and Silvanus
Thompson (1898) pointed to bad syntony as a serious flaw in Marconi's
system; a long review of the system in the *Times* (April 20, 1898) and a lead
article in *The Electrician* (May 13, 1898) agreed with them. Each of these
articles pointed out that heavy damping was a principal cause of bad
syntony.

Marconi could not use Lodge's syntonic principle for two reasons. First,
it was protected by a patent. Second, and more important, the closed transmitter was not a strong radiator. In other words, though Lodge's closed
transmitter was good for tuning, it was bad for practical communication

because it could not transmit far. There was a fundamental and irreconcilable dilemma here: one needed an open circuit to produce powerful radiation, but as a result one ended up with a very "dirty" resonator; one needed a closed circuit to enforce good syntony, but this produced poor radiation. Strong radiation was indispensable for commercial wireless telegraphy, along with good secrecy. Marconi wondered how to secure both characteristics at the same time.

When Marconi initiated extensive experiments at his station at Alum Bay (on the Isle of Wight), in early 1898, the transmitting distance was about 10 miles. Up to that time, he had increased the distance by erecting higher antennas; however, an antenna's mechanical stability became uncertain when its height reached 100 feet. The other means to increase the transmitting distance lay in increasing the sensitivity of the receiver (i.e., the "coherer"). The coherer was activated by a potential difference between its two ends. In Marconi's early receiver, patented in 1896 (figure 4.2), it was attached near the antenna's ground side. The disadvantage of this was that the potential curve along the antenna had a node at the ground, so the potential difference applied to the coherer was very small there. To overcome this, Marconi in effect tried amplifying the voltage in the receiving antenna by connecting the coil's primary windings to the antenna and its secondary windings to the coherer circuit (figure 4.3). After many experiments, he found that with an ordinary induction coil the effect would deteriorate, and that only induction coils made of a very thin wire with a certain numerical relationship between its primary and secondary windings increased the effect. Marconi sensed that this improvement in the sensitivity of the receiver might have some bearing on syntony. When filing a patent on this induction coil (the "jigger," as he later called it), Marconi inserted two sentences concerning syntony in the provisional specification[6]:

It is desirable that the induction coil should be in tune or symphony with the electrical oscillation transmitted.

The capacity of the condenser should be varied (in order to obtain the best effects) if the length of wire is varied.

With this receiver (figure 4.4), Marconi successfully transmitted across the English Channel, a distance of 32 miles.

The receiving jigger was the beginning of Marconi's more efficient syntonic system. In one of his experiments in 1898, he installed a 150-foot transmitting antenna at his station at St. Catherine's on the Isle of Wight

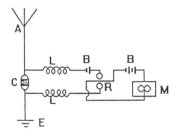

Figure 4.2
Marconi's receiver in 1896. A: antenna. C: coherer. K: capacitance (variable). L: inductance. B: battery. J: jigger. R: relay. M: Morse printer. Source: G. Marconi, "Improvements in Transmitting Electrical Impulses and Signals in Apparatus Therefor," British patent 12,039, filed in 1896.

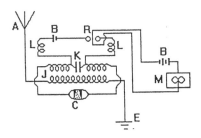

Figure 4.3
Marconi's receiver in 1898. A: antenna. C: coherer. K: capacitance (variable). L: inductance. B: battery. J: jigger. R: relay. M: Morse printer. Source: G. Marconi, "Improvements in Apparatus Employed in Wireless Telegraphy," British patent 12,326 (1898).

and a 56-foot antenna on a ship. At the receiving station at Poole, he erected two receiving antennas with jiggers tuned for each these antennas. The distance between St. Catherine's and Poole was 30 miles, and when the distance between the ship and Poole was 10 miles the two receivers could receive each message separately.[7] From such experiments Marconi discovered empirical rules for his jigger windings that secured the best syntonic conditions.

At that time it was not possible to satisfy the syntonic condition $L_1C_1 = L_2C_2$ by measuring or calculating the capacitance and the inductance of the various circuits, mainly owing to difficulties inherent in the measurement of inductance. Marconi (1901, p. 511) later recalled: "I have found it impracticable by any of the methods with which I am acquainted directly

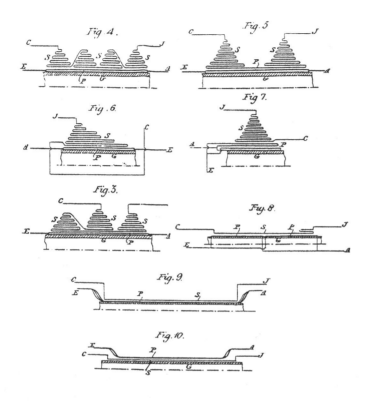

Figure in Drawings.	Diameter of Tube G in centimetres.	Diameter of Wires in centimetres.		Resistance in Ohms.		Number of Turns in Windings.		Length in centimetres.
		Primary.	Secondary.	Primary.	Secondary.	Primary.	Secondary.	
Fig. 3	·935	·01	·01			2 layers of 160 turns each in parallel.	3 sections of 10, 12, 10 layers with 150 45 40 45 40 39 40 35 37 35 30 35 30 25 33 25 20 29 20 15 25 15 12 21 17 5 15 14 10 5 turns—	2.5
4	·937	·012	·012			2 layers of 160 turns each in parallel.	4 sections of 9 layers each with 40 | 80 | 40 35 | 35 | 35 | 35 30 | 30 | 30 | 30 27 | 27 | 27 | 27 23 | 23 | 23 | 23 20 | 20 | 20 | 20 15 | 15 | 15 | 15 10 | 10 | 10 | 10 5 | 5 | 5 | 5 turns—	4.0

Figure 4.4
Marconi's jiggers and syntonic conditions. Source: G. Marconi, "Improvements in Apparatus Employed in Wireless Telegraphy," patent 12,326 (1898).

to measure the inductance of, say, two or three small turns of wire. As for calculating the inductance of the secondary of small transformers, the mutual effect of the vicinity of the other circuit and the effects due to mutual induction greatly complicated the problem." Fleming (1900, p. 90) commented: "Although easy to describe, it requires great dexterity and skill to effect the required tuning." Syntony was not a mathematical principle but a *craft*. Coping with syntony required technical rather than mathematical skills. Marconi, with his jiggers, was the first to master it. The 1898 diary of George Kemp, Marconi's first assistant, shows the stumbling way in which they proceeded:

Nov. 14—No. 26 jigger - 150 primary turns and 200 turns secondary

Nov. 16—Instructed the other station to put on their receiving jigger and they received MM [absolutely] with a detector which gave them Q [good but missing] without the jigger. I then told them to put 5 turns Inductance series with their aerial which gave them stronger signals, but on increasing the number of turns they gave me less energy. Note—Here were points of tuning which required every attention.[8]

Marconi's initial syntonic system was, however, only partially successful. With his receiving jiggers, he was not able to receive two messages at the receiving station "if the two transmitting stations were placed at equal distances from it" or if the signal from one station was much stronger than the other. He reasoned that the weakness in his system lay in the transmitter, and he soon concluded that "some different form of less damped radiator was necessary" (Marconi 1901, p. 509). Reduction of damping was set as a technical goal, but here Marconi was immediately confronted by a fundamental obstacle: whereas a persistent vibrator with less damping (such as Lodge's closed Leyden-jar circuit) was a bad radiator, a good radiator (such as Marconi's straight aerial) was a bad resonator because of its heavy damping. Lodge thought a good resonator to be incompatible with a good radiator.

Faced with this obstacle, Marconi first attempted to insert an inductance coil into the transmitter, as suggested in Lodge's 1898 patent. Through experiments, Marconi determined how inductance influenced signal transmission and tuning. His results were consistent with theory, because, as the damping factor in the L-R-C oscillatory discharge was $R/2L$, either increasing the inductance or decreasing the resistance could diminish the damping. Decreasing the resistance involved shortening the antenna, which jeopardized Marconi's goal of long-distance communication. Increasing the inductance would have been a solution, but in practice adding inductance coils to the transmitter proved unsatisfactory. Inductance was, in a way, like high resistance for high-frequency oscillations. Marconi, therefore, concluded that the failure was caused by the small capacity of conductors in proportion to their inductance (ibid.). Lodge had considered a large capacitance as a kind of reservoir of energy.[9] Max Planck (1897) and S. Lagergren (1898) produced a theoretical formula for a certain oscillator which indicated that large capacitance or large inductance is associated with decreased damping.

To increase capacitance, Marconi first multiplied the number of aerials. That method soon proved impractical, because heavy and high aerials could not be installed on a ship. Next, he made a "concentric cylindrical aerial" for a transmitter and a receiver[10] (figure 4.5). The cylindrical antenna, 3 feet in diameter and 25–30 feet in height, embodied the merits of an open circuit and those of a closed circuit. This was not a true synthesis; the concentric cylinder aerial played the role of a concentric wave guide, and thus radiation emitted from it was very feeble. Though largely experimental, the concentric aerial indicates that Marconi began to regard open and closed circuits

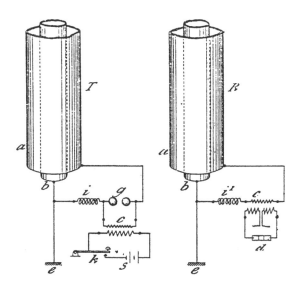

Figure 4.5
Marconi's transmitter and receiver employing concentric cylindrical antenna. This was a "hybrid" between an open and a closed system. Source: G. Marconi, "Improvements in Apparatus for Wireless Telegraphy," British patent 5,387 (1900).

as complementary, not contradictory. Marconi the "practician" was tack-
ling what Lodge and other famous physicists thought irreconcilable.

A revolutionary advance was soon introduced when Marconi adopted a
jigger for the transmitter in order to combine the closed discharge circuit
with the open antenna circuit (figure 4.6). This was described in the "four-
seven" patent,[11] where Marconi made it clear that the antenna circuit
should be suitably attuned for this purpose and specified a dimension of
the transmitting jigger. This method proved extremely good for syntony
between the transmitter and the receiver, as damping was considerably
reduced.[12] Fleming later commented (1903a) that Marconi's patent 7,777
provided "the means of radiating *more or less continuous trains of waves.*"
Having filed this patent, Marconi and Kemp began extensive experiments
to obtain the best syntonic conditions for the transmitting jigger, the transmitting condensers, the receiving jigger, and the receiving condensers. Kemp
recorded some of these experiments in his diary:

April 12, 1900—Exp. bet. Haven & Needle; with 204 = 60ft jigger for reception
at the Needles and five turns of the large impedance in the secondary circuit of our
T jigger, the Needles received M [which meant perfect] and with 30 turns of impedance here the Needle station received S [which meant partially readable], but with
140 turns of impedance here the Needle station received Q [good but missing]. . . .

May 7, 1900—I experimented with 224 = 130 ft jigger also 112 = 110ft jigger here,
and with 202 = 130 ft jigger at the Needles I received M [which meant perfect] with
almost every arrangement that I could make. . . . Some Good Tuning—I experi-

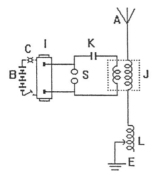

Figure 4.6
Marconi's "four-seven" transmitter in 1900. A: antenna. B: battery. I: induction
coil. C: contact breaker. S: spark gap. K: capacitance. L: inductance. J: jigger.
Source: G. Marconi, "Improvements in Apparatus for Wireless Telegraphy," British
patent 7,777 (1900).

mented with 215 = 150ft jigger at the Needles; 6 jars and 25 turns of "long Henry" here but none at the Needles when this gave that station M, while 6 jars and 30 turns of "long Henry" here gave them no signals. This was good tuning.

May 9—with 112 jigger; 7+4 jars of 30 "long Henry" here and 35 "long Henry" at Needles; 10+4 jars of 49 "long Henry" here and 70 "long Henry" at Needles. Then I tried 202 = 130ft jiggers at the Needles with 7 jars and 29 turns of "long-Henry" when they received M, after which 226 was would with 10ft primary and 10 ft secondary of No. 28 wire on a glass tube 1.4 cm in diameter, when the signals received were weak, missing. . . .

May 14—Marconi left for London[13]

More such experiments were performed by Marconi and Kemp between July and September of 1900. In late September, for the first time, they succeeded in "double reception." Marconi's Niton station sent two messages from two different aerials, and these messages were received separately at the Haven station by two different aerials, each receiving aerial being tuned for each transmitting one. Marconi's syntonic system even increased the transmitting distance considerably. In Marconi's own terms (1901, p. 515), it was a "syntonic apparatus suitable for commercial purpose."[14] The timing was advantageous, since they needed a powerful system for the Poldhu station.

Marconi's new "four-seven" syntonic scheme was not, however, free of problems. Worst of all, it was vulnerable to a patent dispute, since Lodge had also filed a patent on syntony. Ferdinand Braun had also applied (in 1899) for a patent on "Improvements relating to the Transmission of Electric Telegraph Signals without Connecting Wires," which was publicized in January 1900.[15] In his application, Braun, without providing details, described a way of employing a transformer to connect a closed Leyden-jar discharge circuit to an oscillatory antenna. Like Braun, Marconi employed a jigger for inductive coupling. Although Marconi mentioned that the antenna circuit "should preferably be suitably attuned for this purpose," and although he had specified a dimension for the transmitting jigger, Lodge's and Braun's patents rendered the basis of Marconi's 7,777 claim vulnerable to challenge.

Fleming's assistance became crucial at this juncture. His scientific advisorship to the British Edison-Swan Company during the period 1882–1893 had mainly concerned patent matters, and he had acted as an expert witness in other patent disputes. Just before Marconi filed the complete specification of patent 7,777, Fleming, Marconi, John Fletcher Moulton (Marconi's patent attorney), and Major Flood-Page (Managing Director of the

Marconi Company) discussed the best way to fortify the patent's claims. Envisioning litigation with Braun, Fleming asserted that the simplest strategy would be to argue that Braun's oscillatory transformer had itself been preceded by that of Nikola Tesla. But this was clearly not the best tactic, since it would also undermine the basis for Marconi's own claim.[16]

A superior strategy would be to emphasize the novel aspects of Marconi's patent. Fleming singled out three such aspects:

• Marconi's improvement consisted in the tuning of the four circuits in the transmitter and the receiver, requiring the condition $L_1C_1 = L_2C_2 = L_3C_3 = L_4C_4$, where L_i and C_i are the respective inductances and capacitances.
• Marconi's fourfold tuning did, in fact, enormously increase the transmitting distance.
• To produce persistent vibrations in the transmitter, the inductive coupling between oscillator and antenna had to be weak.[17]

Fleming's recommendations were reflected in the complete specification. Partly as a result of this strategy, the patent survived subsequent litigation.[18]

Maskelyne's Attack and the Crucial Marconi-Fleming Demonstration

Marconi's "four-seven" system was adopted in the design of the Poldhu station in 1901. It increased the transmission distance, since it concentrated radiation energy in a small frequency range. Only after Marconi had tuned all the circuits of the Poldhu transmitter with his jiggers did transmission between Poldhu and Crookhaven (a distance of more than 250 miles) become stable.[19] Marconi's "four-seven" system also helped persuade the members of the board of the Marconi Company (who objected to risky experiments on the basis that the powerful waves required for transatlantic signaling would interfere with other ship-to-shore communications, on which the company's success resided) to support the transatlantic experiment. To persuade the board, Marconi had to demonstrate "isolating lines of communication" between the transmitting station at St. Catherine's (the Niton station) and the receiving station at Poole (the Haven station, 30 miles away).

Three experiments were done using Marconi's "four-seven" arrangement. In the first, two transmitters, tuned to specific frequencies, sent different messages at the same time from St. Catherines, and two separate receivers, respectively tuned to these transmitting frequencies, printed two distinct messages. Marconi then connected the two receiving aerials and inductively linked his two receivers to this single aerial. Two messages—one in English, one in French—were then sent from St. Catherines. The two receivers printed these two messages separately despite being connected to a single antenna. In the third experiment, the transmission line between St. Catherines and Poole was crossed obliquely by another line between Portsmouth and Portland. Both signals were transmitted perfectly. These experiments were performed only before Marconi's board, but Fleming later wrote a letter to the *Times* describing the success in commercial tuning.[20]

Marconi's system was not altogether free of difficulties. Syntony proved troublesome during a New York yacht race in the summer of 1901, when Marconi's system and one built by the American inventor Lee de Forest, placed on competing boats to send messages to the press, generated interference with the system of an unknown third competitor.[21] Although this particular event did not have much influence, after Marconi's success in transatlantic wireless telegraphy in December 1901 syntony did become a significant public issue when some individuals who held financial and other interests in submarine cable telegraphy criticized Marconi's scheme. For example, *The Electrician* published the following:

If wireless communication can be established across the Atlantic, it would at all times be perfectly feasible for anyone either in England or in North America—or, indeed, anywhere over a far wider portion of the globe—to erect similar apparatus that would continuously and regularly "tap" every word of every message . . . In fact, if Mr. Marconi establishes wireless telegraphy with America, his signals will be scattered all over Europe as well; . . . everywhere within these boundaries they could be "tapped," and over this entire area they would interfere with other wireless telegraph apparatus working locally.[22]

This kind of criticism addressed both the possibility of tapping (that is, loss of secrecy) and the possibility of interference with other working stations. The first problem also occurred with cable telegraphy and telephony; the second belonged mainly to wireless telegraphy.

On February 20, 1902, Marconi addressed the fifth general meeting of the Marconi Company. After giving a brief description of the past 2 years' experiments on transatlantic telegraphy, and after refuting current skepticism about the accomplishment, Marconi confidently defended his syntonic system, stating that he could transmit across the ocean "without interfering with, or, under ordinary conditions, being interfered with, by any ship working its own wireless installations." He also pointed out that in cable telegraphy an expert could tap messages without cutting the cable. At the end of this speech, he issued a challenge (Marconi 1902a, p. 713):

I leave England on Saturday for Canada and expect to return to England about the end of March. If, on my return, either Sir W.H. Preece or Prof. Oliver Lodge, being of the opinion that, within the space of, say a week, after due notice given to me, he could succeed in intercepting and reading messages to be probably transmitted by me, at stated times, within that period between two of my stations, I should be happy to place any adjacent station of mine at his disposal for the purpose or if he should prefer to conduct his operations from a ship, he would, so far as I am concerned, be quite welcome to do so.

The Electrician immediately responded that the terms of the challenge were not fair, in that "it is not to be expected that either of these experts would be satisfied to use a neighbouring Marconi station, while it is unreasonable to expect them to build for themselves a somewhat costly station and with only a week for the task." *The Electrician* also pointed out that one week would be too short a time (thereby supporting Marconi's statement). But this did not mean that the tapping was impossible. In principle, "syntony would be powerless to prevent any determined attempt to tap the signals which his stations scatter broadcast in all directions." In addition, *The Electrician* argued, showing that tapping was difficult did not address interference at all.[23] As *The Electrician* predicted, neither Lodge nor Preece applied to take up the challenge. It was soon forgotten, buried beneath Marconi's other successes and scandals.

In March 1902, during a trip to the United States, Marconi received "SSS" messages with a Morse printer across 1500 miles. This success was followed by harsh criticism from Silvanus Thompson, who revealed that neither the transmitting station at Poldhu nor the mercury coherer used at St. John's was of Marconi's own design. Indeed, Thompson claimed that the transmitter had been designed by Fleming and that the coherer had been invented by the Italian engineer Luigi Solari. Thompson argued further (and not for the first time) that Oliver Lodge was the true inventor of wireless telegraphy (Thompson 1902a; see also chapter 2 above).

Shortly after Thompson's attack, Marconi's attempt to file his own patent on Solari's coherer was revealed. This raised questions concerning his ethics. Marconi escaped the questions by announcing the invention of a new magnetic detector based on Ernest Rutherford's 1896 discovery of the magnetizing effect of electromagnetic waves. The detector freed Marconi from the damaging effects of the Solari's coherer.[24] From July to September 1902, Marconi continued experimenting on long-distance wireless telegraphy between Poldhu and his laboratory aboard the *Carlo Alberto*. While sailing to Italy in September, Marconi succeeded in receiving ordinary telegraphic messages (not merely "SSS") from Poldhu at a distance of 750 miles across both land and sea. This news, along with news of the first long-distance telegraphic messages, was widely publicized by Solari's report in *The Electrician* (1902).

Even the skeptical *Electrician* admitted that Mr. Marconi was to be congratulated.[25] However, *The Electrician* stated that the signals had been tapped in England "with instruments not tuned by the Marconi Company for the purpose." Marconi's experiments, the article maintained, had in fact demonstrated both the lack of secrecy in long-distance wireless telegraphy and the possibility of tapping at other stations.[26] The Managing Director of Marconi's Company quickly denied *The Electrician*'s claim, but the journal then published a detailed report on tapping by Nevil Maskelyne. Maskelyne published his Morse papers on which dots and dashes were printed and claimed that the messages printed on his papers were exactly the same as those Solari had received aboard the *Carlo Alberto* sailing to Italy. And Maskelyne (1902) made other remarks which further startled Marconi and his company:

I had been almost constantly in touch with Marconi stations in various parts of the country. In every case I have found that our working produced mutual interference. Consequently, I inferred that those particular stations were not fitted with Mr. Marconi's syntonic apparatus. The reason for this I could not guess; but such was the fact. Of course, I had read of the marvelous efficacy of syntony. I had read of "triumphs" achieved almost every other day. I had read of experimentalists—even Sir Oliver Lodge, to whom we all owe so much—being challenged to intercept Marconi messages. Then the question arose, why does not Mr. Marconi use this syntonic apparatus? It seemed to be something too precious to be supplied to mere working stations. Still, he must use it *somewhere,* and the only conclusion at which I could arrive was that the syntonic arrangements must be employed exclusively at Mr. Marconi's latest and greatest station at Poldhu. Yet, when I went to Porthcurnow, 18 miles distant, I received Marconi messages with a 25 ft. collecting circuit raised on a scaffold-pole. No wonder I was interested. When, eventually, the mast was erected and a full-sized collecting circuit installed, the problem presented was, not how to intercept the Poldhu messages, but how to deal with their enormous excess of energy. That, of course, involved no difficulty, and by relaying my receiving instrument through landlines to the station in the valley below, I had all the Poldhu signals brought home to me at any hour of the night or day. It is for this reason that I claim to know something of the experiments conducted between Poldhu and the *"Carlo Alberto."*

If Marconi's syntonic messages were so easy to intercept, Maskelyne asked, "what has become of that syntony of which we have heard so much?" Maskelyne (ibid.) now challenged to Marconi: "Can Mr. Marconi so tune his Poldhu station that, working every day and all day, it does not affect the station at Porthcurnow? . . . he had only succeeded in proving that he cannot do so."

Maskelyne, who came from a wealthy and well-known family, was a self-educated electrician who had been interested in wireless telegraphy since the late 1890s. His demonstration in 1899 that gunpowder could be exploded by wireless control generated widespread public interest. In 1900, at the annual meeting of the British Association for the Advancement of Science in Bradford, he demonstrated a system for communication between a balloon and a base separated by about 10 miles.[27] With H. M. Hozier, secretary of Lloyd's of London, he had attempted to develop a wireless telegraphic system for the insurance company; however, without employing a grounded antenna (that is, without violating Marconi's patent) they could not obtain practical results. After Lloyd's contracted with Marconi for that purpose, Maskelyne began to criticize Marconi's monopoly of wireless telegraphy. Eventually he became a leading figure in the anti-Marconi faction. Maskelyne had several patents on wireless telegraphy, but after his debacle with Lloyd's he seemed to be more involved in attacking Marconi than in developing his own practical system.[28]

After the publication of Maskelyne's report, critical opinions concerning Marconi's Poldhu experiments poured forth. It was argued in *The Electrician* that "it is far more important that we should possess an effective ship-to-ship and ship-to-shore telegraphic system than that the Marconi Company should be able to establish telegraphic communication across the Atlantic."[29] The Marconi Company castigated Maskelyne's printed messages as forgeries and asserted that tapping by an expert was not the issue, since it was already possible in cable telegraphy and in telephony. Instead, the company argued that its priority was to develop transmission technology that would not interfere with ordinary ship-to-shore communication, that it was confident in its ability to do so, and that it in fact had done so at Poldhu (Hall 1902). In short, the company began to emphasize interference rather than tapping as the major problem. Critics requested definite "proof of the justice of their repeated claim that the Poldhu station can be worked . . . without interfering with other stations in its neighbourhood."[30]

On February 9, 1903, Marconi returned to England from the United States. While in the United States, he had helped President Theodore Roosevelt send a message to King Edward by wireless telegraphy. However, the king's reply came by cable telegraphy, because a post office near Poldhu, which had been closed at the time, had not dispatched the king's telegram to the Poldhu station. Marconi blamed the Post Office for this neglect. At the same time, the Anglo American Telegraph Company, which was in charge of the business of transatlantic submarine telegraphy, publicly criticized Marconi's wireless telegraphy for its lack of secrecy. In response to this criticism, Marconi announced in an interview to the *St. James Gazette* (published February 9) that he was "quite prepared to accept a licence subject to revocation if the naval installations or the existing land systems were interfered with through induction." Marconi also stated that "during the experiments on the *Carlo Alberto* an installation was set up close to our station and tapping did take place, but then no attempt was made at secrecy." Maskelyne then wrote an angry letter to the *St. James Gazette* pointing out that his station at Porthcurnow was 18 miles away, and that he had to make efforts to obtain secrecy, rather than to tap messages.[31]

Marconi decided to design a crucial "show" of the workability of his syntonic system and of the safety of the Poldhu power station (figure 4.7). The program was prepared by Fleming. A small transmitting station simulating an ordinary shipboard station was built 100 meters from the gigantic Poldhu station. The power of the small station was $\frac{1}{100}$ that of the large one. The two stations were tuned to different frequencies. Two of Marconi's duplex receivers, one tuned to the small station and the other to the large station, were installed at the Lizard station. The experiment consisted in transmitting two different messages from these two stations at the same time and receiving them separately at the Lizard station.

Fleming prepared sixteen messages. Eight were typical ship-to-shore messages for the small station; the other eight included some cipher messages in ABC and some simple English sentences. These were to be transmitted by the big Poldhu station. Keeping copies for later comparison with the transmitted signals, Fleming put the messages into sealed envelopes for secrecy, without showing them to Marconi. He inspected the Poldhu station to determine that it was working to its full capacity. To be certain that it was, Marconi arranged for another of his stations, this one located 200 miles away at Poole, simultaneously to receive the messages sent from the Poldhu

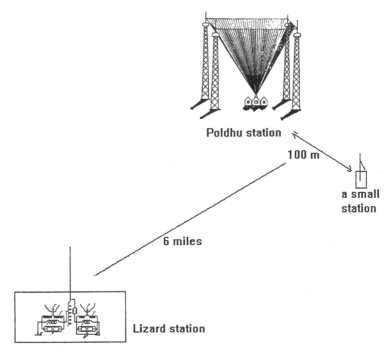

Poldhu station

100 m

a small
station

6 miles

Lizard station

Figure 4.7
The crucial Marconi-Fleming demonstration. Two messages sent separately from
the "big" Poldhu station and a nearby small station were captured separately at the
Lizard station, 6 miles away.

station. Since Fleming was to inspect the messages received at the Lizard in
the company of Marconi, he arranged for a disinterested assistant, "unconnected with the Marconi organization," to remain at Poldhu, where his
duty was to open the envelopes at a certain time and to deliver the first two
messages to the operators of the two stations. The operators were to send
these messages for 10 minutes; then, after a 5-minute interval, the assistant
was to give them the envelopes containing the second two messages, and so
on. The experiment was conducted on March 18, 1903, from 2 to 4 P.M.
Marconi and Fleming watched the messages print at the Lizard station.
Afterward, Fleming asked Marconi to read the messages that they thought
they had received from the small station. Marconi read off and wrote down
the eight messages "without a single mistake."

In his Cantor Lecture of March 23 (Fleming 1903a, p. 772) and in a long
letter to the *Times* dated April 4,[32] Fleming proudly claimed that these

results were a satisfactory demonstration of the lack of interference between Marconi's syntonic systems. At the sixth general meeting of the company, held on March 31, 1903, Marconi (1903b) declared that Fleming's experiments confirmed his belief that no interference would occur between the Poldhu station and ordinary ship-to-shore communications, provided that they were all equipped with Marconi Company equipment. However, Marconi's and Fleming's triumph did not go long unchallenged.

Maskelyne Interferes with Fleming's Royal Institution Lecture

The Marconi-Fleming experiment had not been a public demonstration, nor had an unimpeachable witness inspected the received messages (which, after all, had been read by Marconi and compared with copies kept by Fleming). The authenticity of the demonstration, and thus of the syntony and safety of Marconi's system, relied entirely on Fleming's own testimony.[33] His authority resulted from the reputation he had accrued over the previous 20 years within the British electrical engineering and physics community. Fleming was known as a meticulous, honest, hard-working scientist-engineer who had laid the foundation of scientific electrical engineering (Hong 1995a,b).

But was Fleming's reputation valued in the new world of wireless telegraphy? Physicists like Oliver Lodge, power engineers like Fleming, telegraphers like Alexander Muirhead and William Preece, and self-educated inventors like Marconi and Maskelyne were now competing for authority in the novel world of "ether engineering." Furthermore, as a result of Marconi's own virtual monopoly of the market, the challenge to him was intense and came from several directions. As a result, from 1899 to 1903 Fleming had had to work hard to gain credibility in this new field, and he had done so almost entirely by translating Marconi's practical wireless telegraphy into the language of scientific engineering. His two Cantor Lectures (Fleming 1900, 1903a) were read worldwide and were cited as providing the first scientific rationale for wireless technologies. Fleming had now used his authority as a qualified witness to publicize Marconi's secret demonstrations. To Marconi's opponents, therefore, challenging Fleming's credibility became essential.

Just after Fleming announced Marconi's success in practical syntony, Maskelyne claimed "a right to demand the absolute justification of

[Fleming's] claims."[34] Several days later, at a company meeting, Marconi (1903b) stated that Fleming would repeat the experiment under the supervision of Lord Kelvin and Lord Rayleigh. This repetition was not performed, however. Instead, Fleming prepared laboratory-scale demonstrations to be given at the Royal Institution in a pair of public lectures on "Electric Resonance and Wireless Telegraphy." The first lecture, given May 28, 1903, mainly concerned theoretical aspects of Hertzian resonance and their practical implications for wireless telegraphy.[35] In the second lecture, given June 4, Fleming discussed conditions for tuning and for the generation and reception of syntonic oscillations. According to the brief summary that appeared 2 weeks later in *The Electrician*,[36] Fleming had "referred to the possibility of tuning transmitters and receivers . . . , pointing as an instance, to the experiments recently carried out by Mr. Marconi on the south coast of England." Then, *The Electrician* continued, Fleming had demonstrated the reception of two wireless messages, one sent from Fleming's laboratory at University College London and the other from Marconi himself at the distant station in Poldhu. These messages, it was noted, had not been received directly at the Royal Institution; they had received at Chelmsford (where Marconi had his factory) and then re-transmitted; the reason given for this contrivance was that the antenna erected temporarily at the Royal Institution was simply too short (60 feet) to receive the Poldhu message directly (figure 4.8). These demonstrations were not designed to show lack of interference between different systems transmitting simultaneously at different frequencies, but rather to demonstrate the power of wireless signaling with a tuned system. But Maskelyne had other goals in mind.

According to his recollection, Maskelyne at first planned to attend the lectures but "at once grasped the fact that the opportunity was too good to be missed." He decided to test Marconi's claim of non-interference by trying to interfere in Fleming's lecture on syntony. This, he felt, was "something more than a right; it was a duty." Marconi had always claimed that even the gigantic Poldhu station would not interfere with his low-power ship-to-shore communication. If this was really the case, Maskelyne reasoned, ordinary transmitting stations should certainly not trouble communications between Marconi stations. He installed a 10-inch induction coil at the Egyptian Theatre (which his father owned), near the Royal Institution, and adjusted his transmitter so that the radiation was "out of tune" with Marconi's. He used short waves, as Marconi was known to use

Figure 4.8
The antenna erected on the roof of the Royal Institution for Fleming's lecture with
which Nevil Maskelyne interfered on June 4, 1903. Source: *Wireless Telegraphy*,
ed. E. Sharman (S.C. Mss. 17, Institute of Electrical Engineers).

long waves. The final problem was how to check whether his trial was successful or not. For this purpose, he sent messages "calculated to anger and draw somebody at the receiving end."[37] The lecture began at 5 P.M. Around 5:45 Maskelyne began transmitting.

Thanks to Arthur Blok (1954), who assisted Fleming at the lecture, we have a detailed description of what happened in the lecture theatre of the Royal Institution:

One of the Marconi Company's staff was waiting at the Morse printer and while I [Arthur Blok] busied myself with demonstrating the various experiments I heard an orderly ticking in the arc lamp of the noble brass projection lantern which used to dominate this theatre like a brazen lighthouse. It was clear that signals were being picked up by the arc and we assumed that the men at Chelmsford were doing some last-minute tuning-up.

But when I plainly heard the astounding word "rats"[38] spelt out in Morse the matter took on a new aspect. And when this irrelevant word was repeated, suspicion gave place to fear. The man at the printer switched on his instrument and the hands of the clock moved inexorably towards the minute when the Chelmsford message was to come through. There was but a short time to go and the "rats" on the coiling paper tape unbelievably gave place to a fantastic doggerel, which, as far as my memory serves me, ran something like this

There was a young fellow of Italy
Who diddled the public quite prettily

A few more lines *en suite* completed the verse, and quotations from Shakesphere also came through.

Evidently something had gone wrong. Was it practical joke or were they drunk at Chelmsford? Or was it even scientific sabotage? Fleming's deafness kept him in merciful oblivion and he calmly lectured on and on. And the hands of the clock, with equal detachment, also moved on, while I, with a furiously divided attention, glanced around the audience to see if anybody else had noticed these astonishing messages. All seemed well—a testimony to the spell of Fleming's lecture—until my harassed eye encountered a face of supernatural innocence and then the mystery was solved. The face was that of a man [Dr. Horace Manders] whom I knew to be associated with the late Mr. Nevil Maskelyne in some of his scientific work.

The story ended happily, at least as far as the lecture was concerned. By a margin of seconds before the appointed Chelmsford moment, the vagrant signals ceased and with such *sang froid* as I could muster I tore off the tape with this preposterous dots and dashes, rolled it up, and with a pretence of throwing it away, I put it in my pocket. The message from Chelmsford followed and Fleming's luck had held amidst much unsuspecting applause.[39]

Three points in this recollection should be noted. First, it appears that neither Fleming nor Marconi had the slightest expectation of any attempt to ruin their public show.[40] Fleming designed his public demonstrations with the utmost care, and he was confident in the efficacy of Marconi's system.

Second, Marconi's antennas and receivers at the Royal Institution were supposed not to pick up any wireless messages other than Marconi's own, because they were designed on the basis of Marconi's "four-seven" system. Third, Fleming received the Poldhu messages (relayed by Chelmsford), not because of a lack of interference from Maskelyne's messages, but because the latter's messages had stopped before the ones from Chelmsford arrived. Had Maskelyne's messages continued for a few more minutes, they would have certainly generated jumbled results by mixing with Marconi's.

Fleming, told about the attempt after the lecture, was infuriated. On June 5 he reported to Marconi:

Everything went off well at the Royal Institution and we got through your reply to the President's telegram both before and at the lecture perfectly, and I have the tapes this morning from Chelmsford. There was however a dastardly attempt to jamb us; though where it came from I cannot say. I was told that Maskelyne's assistant was at the lecture and sat near the receiver. Nothing of the kind happened at previous rehearsals but at 5:45 Mr Woodward [an engineer working at the Marconi Company] got a message down which came "through the ether" from somewhere and there were mysterious effects heard in the arc lamp which seemed to indicate "electric jerks" as Lodge would say being put with the earth plate or the aerial. Maskelyne's assistant had been seen loafing about some days before. Anyway the attempt was not successful in spoiling our "show." We got of course excellent signals from my laboratory. I will tell you more about this attempt to jamb us when we meet, but I want to investigate it first, and if I can find out who did it it will not be pleasant for him.[41]

We do not know whether Marconi welcomed Fleming's intention to investigate, since Marconi's reply has not survived; however, we do have Fleming's next letter to Marconi, in which he stated that Maskelyne had interfered with the demonstration by flowing a strong earth current nearby. "Professor Dewar [the director of the Royal Institution] thinks I ought to expose it," Fleming wrote. "As it was a purely scientific experiment for the benefit of the R.I. it was a ruffianly act to attempt to upset it, and quite outside the '*rules of the game*'. If the enemy will try that on at the R.I. they will stick at nothing and it might be well to let them know."[42] Why was it outside the rules of the game? Fleming seemed to think that "tapping" of wireless messages was within the rules of the game. Fleming apparently believed that someone had created a strong earth current to destroy the grounding of Marconi's antenna, thereby interfering with syntonic communication in an unacceptable fashion. In other words, he had complete confidence in the efficacy of Marconi's "four-seven" system. Yet, contrary to Fleming's conjecture,

Maskelyne had not in fact used an earth current; he had used electromagnetic waves.[43] Fleming's mistake lay in assuming that if Maskelyne *had* actually sent out proper air waves he would have used a syntonic transmitting system similar to the one Fleming and Marconi were employing. In other words, anyone seeking to interfere with Marconi's syntonic system would have used a narrowly defined frequency different from the one Marconi was using. Marconi's "four-seven" devices, Fleming was utterly convinced, were protected from interfering with one another; thus, if Maskelyne had succeeded, it could not have been by using air waves. It apparently never occurred to Fleming that Maskelyne would use a simple induction coil to generate short waves. Maskelyne, perhaps unintentionally, had exploited "dirty" waves (waves having a broad frequency range), to which Marconi's syntonic system was vulnerable. The old, simple technology of the spark-gap transmitter, which Fleming and Marconi, along with many others, had long ago thrown into the trash bin, had now trumped them. Fleming could not imagine Maskelyne to have used such an inherently faulty and outdated technology, and so he concluded that Maskelyne must have deliberately tricked him in the simplest way possible: by destroying the efficacy of Fleming's grounded antennas by sending a strong current directly to earth from a nearby hiding place. Maskelyne, for his part, probably did not intentionally employ a device that he knew ahead of time would be detected by any device whatsoever, no matter how it was designed. Indeed, he seems not fully to have understood the reason for his success in interfering with Marconi's syntonic system.

Who Was the Hooligan?

Initially, Maskelyne's interference was kept secret. Only Maskelyne and Fleming, their assistants, Marconi, and a few of Marconi's employees knew of it. But on June 8 Fleming wrote a letter to the *Times* (published on June 11) in which he stated that a "deliberate attempt" to "wreck the exhibition" of London-Poldhu communication had been undertaken by "a skilled telegraphist and someone acquainted with the working of wireless telegraphy." Fleming then emphasized the scientific nature and purposes of his own demonstration, invoking the sacred name of Faraday and bemoaning that "the theatre which has been the site of the most brilliant lecture demonstrations for a century past" was not protected from the "the attacks of a *scientific hooliganism.*" He then begged the readers for the names of those responsible for the "monkeyish pranks."[44] The *Daily Telegraph* immediately interviewed Fleming, and Cuthbert Hall, the managing director of the Marconi Company. Fleming stated that "the public opinion would have condoned an attempt to make the perpetrators themselves the subject of 'a striking experiment.'" Hall stated that the apparatus Fleming used at the Royal Institution was not as stable as that in Marconi's commercial stations, having been fabricated rather quickly for the lecture, and commented "we can afford to laugh at it" since the Poldhu messages had gotten through.[45]

In a letter to the *Times*, Nevil Maskelyne, who was anxiously waiting for Fleming's response, frankly and proudly admitted what he and his friend Horace Manders had done. However, Maskelyne denied Fleming's allegation that he had attempted to "wreck" the show. He defended himself by asserting that he could have easily done so had he wanted to. In any case, he said, he had transmitted only for a few minutes. He justified his behavior on two grounds: (1) Fleming had been attempting to demonstrate not the scientific principles of Hertzian syntony, which were not in question, but "the reliability and efficacy of Marconi syntony," which were. (2) Maskelyne had adopted "the only possible means of ascertaining *fact* which ought to be in the possession of public." What was this fact? It was that "a simple untuned radiator upsets the 'tuned' Marconi receivers."[46]

After Maskelyne's confession, the *Daily Express* interviewed Fleming. Evidently taking some pleasure in having the perpetrator identified for him, Fleming remarked that he "had anticipated some attempts of the kind" and had taken "some suitable precautions." Ignoring or overlooking Maskelyne's reference to his "simple untuned radiator," Fleming repeated his belief that "there was clear proof that somebody quite close was putting high-tension currents into the earth," and that this made the attempt "not a fair interference" but "getting in at the back door." In spite of all this, "there was no actual success in interfering with my lecture." Finally, Fleming again referred to how a "scientific lecture" at the "sacred" Royal Institution "should be kept free from things of this sort."[47] Fleming, however, avoided answering questions concerning the efficacy of Marconi's syntonic system. When asked, he just repeated that the interference was unfair and proved nothing.

Fleming invoked Science and the Royal Institution in blaming Maskelyne, whereas Maskelyne defended himself in the name of the public. Fleming's claim depended to a considerable extent on the implicit argument that

Fleming had not been dealing specifically with Marconi's syntony, and that he had been speaking about physical principles that transcended the particularities of the commercial world. As the dispute continued, Fleming seems to have adopted a strategy of separating his "scientific" lecture from Marconi's "commercial" syntonic wireless system.[48] Such a strategy might protect both Fleming and Marconi. It could be said that Fleming's instruments were not of a commercial sort, and that Maskelyne's "monkish prank" ruined only Fleming's scientific lecture and not Marconi's commercial system. To Fleming, who thought of himself as a scientific man, this distinction between science and commerce was natural. To the public, however, Fleming was inseparable from Marconi. His argument probably was no more compelling to many others than it was to Maskelyne. The receiving antenna and the other apparatus in the Royal Institution had been installed with Marconi's help, and it was Marconi who had sent messages from Poldhu for Fleming's lecture. Further, Fleming had definitely stated in the lecture, as one observer later recalled, that "his instruments were so arranged that no other person could prevent messages from Poldhu from reaching him" (Black 1903).

Maskelyne replied: "The Professor called up the name of Faraday in condemning us for what we did. Supposing Faraday had been alive, whom would he have accused of disgracing the Royal Institution—those who were endeavouring to ascertain the truth or those who were using it for trade purposes? Professor Fleming gave two lectures that afternoon. The first was by Professor Fleming, the scientist, and was everything that a scientific lecture ought to be; the second was by Professor Fleming, the expert advisor to the Marconi Company."[49] To many, Fleming's name was by then so intimately connected to that of Marconi that it was difficult to associate him with pure Hertzian physics, and in any case it was commercial telegraphy, not the rarefied realm of "ether waves," that bore on public interests. In a letter to the *Times*, Charles Bright, an old and respectable telegrapher, requested a "full, open, and disinterested inquiry as to the merits of the system from every standpoint, accompanied by a complete series of trials by impartial authorities."[50] In an editorial titled "Who Was the Hooligan?" the *Electrical Review* sided with Maskelyne:

If it turns out that Mr. Maskelyne made use of extraordinary means to upset the lecturer, Prof. Fleming has some grounds for his protest; but if, as we think, the means employed were fair, and such as might occur in practice, then the protest must be made by the public against the professor. In his dual capacity of *savant* at a learned institution, and expert to a commercial undertaking, Prof. Fleming is discovering that while the public have nothing but courtesy and respect to offer to the one, they have searching criticism and still more searching experiment to oppose, if need be, to the statements of the other. When the philosopher stoops to commerce he must accept the conditions of commerce. Having descended from the high pedestal, upon which science placed him, into the arena of trade and competition, he attempts, when attacked, to climb back to the sacred temple, whence, like the proverbial armadillo, he can in safety evade, observe and deride his pursuers. But, as a matter of fact, it is ridiculous for him to appeal in his dilemma to the reverence which we all feel for the traditions of the Royal Institution. Faraday, to begin with, would never have placed himself in so anomalous a position; and, moreover, Faraday would have displayed more interest in what Mr. Nevil Maskelyne was trying to uncover, than in what, if we are to believe Mr. Maskelyne's statement, Prof. Fleming was seeking, for commercial reasons, to withhold from public knowledge.[51]

The controversy ended in an unexpected way. In addition to Fleming's second lecture, a few more demonstrations were scheduled for June 4. In the morning, Marconi was to send a message from Poldhu to Professor Dewar at the Royal Institution via Chelmsford. That message was to be captured at the Royal Institution, inspected by the president of the Royal Society, then passed to Dewar. Dewar was then to dispatch a congratulatory telegram (by cable) at 3 P.M., which Marconi at Poldhu would easily receive by 4 P.M. Finally, Marconi was to send a wireless reply to Dewar via Chelmsford during Fleming's public lecture; it was to be read to the audience by Fleming. Thus, the Poldhu station was to have sent two different messages from Marconi: his first message, in the morning, and his reply to Dewar, after 4 P.M.

However, according to the *Morning Advertiser*, which got its information from Maskelyne, the Poldhu station sent only one message in the period between 11:50 A.M. and 4:30 P.M.: "BRBR. Best Regards to Prof. Dewar sent through ether from Poldhu. —Marconi. PDPD," This must have been Marconi's first message to Dewar; it could not have been his reply, which was (again according to the *Morning Advertiser*) "My best thanks to the President Royal Society and yourself, for kind message; Communication from Canada was re-established May 3rd." The doubt raised by Maskelyne concerned why Marconi's first message was sent from Poldhu repeatedly until 4:30 P.M. When did Marconi receive Dewar's telegram? If Marconi had not received Dewar's telegram by 4:30 P.M., had Marconi's first message not reached the Royal Institution? When a reporter for the *Morning Advertiser* asked Fleming for his opinion, Fleming's reply was: "I have nothing to say; nothing further on my account."[52] Fleming's reply may have amplified the doubts. Maskelyne (1903b) raised the issue again in the July 10 issue of *The Electrician.*

In its July 17 issue, *The Electrician* published a letter from Maskelyne and a letter from the Marconi Company replying to Maskelyne's charge. In his letter, Maskelyne admitted that he had been wrong. It had been found that another station, 50 miles from Poldhu, had captured Marconi's second message repeatedly sent out from the Poldhu station (Marconi's reply to Dewar concerning the re-establishment of transatlantic communication) after 5 P.M.—that is, during Fleming's lecture (Maskelyne 1903c). The Marconi Company replied to Maskelyne's attack of July 10 with a detailed record of operations at the Poldhu and Chelmsford stations on June 4, and with a copy of the message sent from both stations; the latter was signed by two technicians from each of these two stations and by Fleming and his assistant at the Royal Institution. The company's rebuttal certified that the message, which read "My best thanks to President Royal Society and yourself for kind message; Communication from Canada was re-established May 23rd" (just as printed in *Morning Advertiser*), had indeed been sent from Poldhu to Chelmsford at 5:15 P.M. and relayed from there to the Royal Institution at 5:25 (Allen 1903).

Unfortunately for Fleming and Marconi, the specific message that the company certified to have been sent out from Poldhu did not tally well with the one received at the Royal Institution. According to Maskelyne (1903d), the message read at the Royal Institution by Fleming specifically ran "P.D. to R.I. - To Prof. Dewar. To President Royal Society and yourself thanks for kind message; Communication from Canada was re-established May 23rd —Marconi." The messages clearly had the same meanings, but, as Maskelyne had said, the wording was somewhat different. Fleming had, no doubt, received some message from Chelmsford, but it was apparently not the same message that the Marconi Company had sent out from Poldhu.

In view of the evidence that the message sent out from Poldhu was indeed captured at another station nearby, there remain two possibilities: (1) Technicians at Chelmsford, while dispatching the message, may have somehow mistranscribed the message from Poldhu. (2) The message sent from Poldhu had not in fact been received, or had been very imperfectly received, at Chelmsford. In the latter case, the message Fleming received at the Royal Institution must have originated at Chelmsford with someone who had somehow known or had simply guessed at the content of

Marconi's congratulatory message. Imperfect reception of messages was rather common at that time.

Maskelyne had few doubts about what had probably happened. As he put it, "Chelmsford could easily have sent such a message without having received a word from Poldhu." After this revelation, the Fleming-Marconi side stopped responding. The controversy ended with doubt having been cast on the authenticity of Fleming's entire scientific show. *Punch* published a lampoon about a daily newspaper on an ocean liner produced by means of "Marconigrams"; the joke was that the stories were topsy-turvy because of interference from a nearby ship.[53]

The Maskelyne affair did considerable damage to Fleming's credibility. He could no longer claim to be a "qualified witness" where Marconi was concerned. He could not, and did not, write any further letters to the *Times* about Marconi's secret demonstrations.

The affair forced Marconi to moderate his love of public demonstrations. Since 1896 he had used public demonstrations to attract attention and to exhibit the practical nature of his wireless telegraphy. Such demonstrations had been performed for and by famous people, including Lord Kelvin, Alfred Lord Tennyson, Prince Edward (later King Edward VII) and other members of the British Royal Family, the king of Italy, the Russian emperor, and the president of the United States. Reports of these performances published by the Marconi Company glowingly described success after success. Yet some of the actual performances were rather different from the company's portrayals of them. In September 1902, for example, Marconi had been keen to receive on board the *Carlo Alberto* congratulatory messages sent from Poldhu for the Italian king. Solari's published report of the event stated that this had gone quite smoothly. In fact, however, reception had been so poor that the temperamental Marconi had smashed all the receivers. During his second round of transatlantic experiments in the winter of 1902–03, transmission and reception had been difficult and unstable, though the published papers described the results as if they had been entirely straightforward (D. Marconi 1982, pp. 116–123).

The Cycle of Credibility

The Maskelyne affair illuminates several interesting aspects of the early history of wireless telegraphy. First, it betrays a major, indeed critical, limitation of Marconi's syntonic system. Any syntonic system was extremely vulnerable to interference from old, simple transmitters that generated "dirty" waves. In 1901, Marconi had declared "the days of the non-tuned system . . . numbered" (Marconi 1901, p. 515), and Fleming had confidently remarked in *The Electrician* (May 24) "It will now be possible to provide the Admiralty and the Post office with instruments having an Admiralty or a Post Office frequency, and to register frequency just as a telegraphic address is registered, so that no one else could use that particular frequency." The Maskelyne affair pointed out that avoiding interference would require a sort of instrumental monopoly under which unsyntonized transmitters would be strictly prohibited. The affair reflected, and accelerated, wireless technology's transition from a period in which public shows and sensations (such as the first transatlantic reception of SSS signals) had been essential to economic success to a period in which regulating frequencies and guaranteeing instrumental uniformity became serious issues.

The Maskelyne affair had a major impact on J. A. Fleming's professional career. In 1899, Fleming had been appointed scientific advisor to the Marconi Company for a year, at an annual salary of £300.[54] As was noted in the previous chapter, in December 1900, when he was fully engaged with Marconi's first efforts at transatlantic telegraphy, Fleming succeeded in getting a raise to £500 a year and in getting his engagement extended for 3 years. From 1899 to 1903, and particularly after transatlantic wireless telegraphy was first achieved, Fleming's role was largely limited to helping Marconi create public demonstrations and reporting Marconi's private shows to the British public. The Maskelyne affair strikingly undermined his credibility, preventing him from serving as a trusted witness. Moreover, his contract with the Marconi Company, which terminated in December 1903, was not renewed. He had become useless to Marconi.

Fleming returned to the Pender Laboratory at University College, where in 1904 he invented a wavelength-measuring instrument (the "cymometer") and a high-frequency AC rectifier (the "thermionic valve"). In May 1905, with these devices in hand, he reestablished his relationship with Marconi, having created new technical capital and, in direct consequence, a new source of credibility.

5

Transforming an Effect into an Artifact:
The Thermionic Valve

I have not mentioned this to anyone yet as it may become very useful.
—John Ambrose Fleming to Guglielmo Marconi, November 1904 (Marconi
Company Archives)

Claiming the Valve: Invented, Improved, and Litigated

Almost all home radios built between the 1920s and the 1960s employed
glowing "lamps" in their receivers. In the eyes of non-experts, these resembled Thomas Edison's incandescent lamps. Of course, they were actually
vacuum tubes for amplifying signals. Vacuum tubes were also used in transmitters, to produce continuous waves. Their ancestor was Fleming's
thermionic valve, invented in 1904. The thermionic valve—the first tube to
contain an extra electrode—was used for wireless reception in the field. It
rectified the high-frequency alternating currents that are induced in a receiving antenna by electromagnetic oscillations.

In 1906, inspired by Fleming's thermionic valve, Lee de Forest invented
the audion, one of whose three electrodes acted as a control grid. Beginning
in the early 1910s, de Forest's audion (or triode) was used not only as a rectifier but also as an amplifier and as an oscillator. This opened the period
of radio broadcasting.

The story of the invention of the thermionic valve is well known, partly
because the valve was the "stem cell" for all the various vacuum tubes that
were developed throughout the twentieth century, but also because its
invention was dramatic. The popular story goes as follows: A curious
effect—"the Edison effect"—was accidentally discovered in the early 1880s
by Thomas Edison and his electricians, who did not understand its mechanism. John Ambrose Fleming, a physicist-engineer working for the British

Edison Company, investigated the effect in the 1880s and the 1890s; he understood it in terms of "unilateral conductivity" inside the lamps. While serving as scientific advisor to Marconi in 1904, Fleming used his understanding of the process of unilateral conductivity to create the valve, which he designed as a signal detector for wireless telegraphy. In this story, Fleming mediates between physics and engineering, and between the area of engineering associated with the lamp and the area associated with the radio. This mediation was not haphazard, the story goes, because Fleming had been working on the Edison effect longer time than anyone else; in 1904 this had led him to seize a momentary opportunity that others might well have overlooked.[1] This story—first told by Fleming himself in the late 1910s, while he and the Marconi Company were fighting with Lee de Forest over the patent on the vacuum tube—is misleading.

When the "authorship" of a technology is at stake in court, recalling how one invented or contributed to the invention of the technology is important. Different strategies are used. An inventor may point to radical differences between his technology and others. The inventor may emphasize the continuity between his past work and the new technology in question. Simultaneous discontinuity with others' work and continuity with his own earlier work often characterize an engineer's recollections of the invention of a novel artifact. All these factors are present in Fleming's recollection of the invention of the thermionic valve.

Fleming's story is interesting for several reasons. Above all, the Edison effect was well known in the 1880s and the 1890s, and no significant changes, it seems, were involved in the transformation of the Edison effect into the technology of the thermionic valve. So why was it Fleming who carried this through? Could someone else have easily done the same? Was Fleming merely lucky? Fleming would have argued that, although the Edison effect was well known, few scientists and engineers experimented with it, and even fewer became interested in wireless telegraphy after 1896. Therefore, Fleming may have been the only individual to have performed extensive experiments on the Edison effect in the 1880s and the 1890s and to have then worked on wireless telegraphy. But why were his experiments of the 1880s and the 1890s so essential, if, as the popular story implies, the transformation of the Edison effect into the thermionic valve was smooth and even inevitable?

Fleming overemphasized, or even distorted, the smoothness of the transition from an effect to a device, the extent of his research on the effect, the importance of his scientific advisorship to Marconi as a context for the invention, and the notion that he envisioned a well-defined use for the thermionic valve as a detector in wireless telegraphy when he first invented it, and all these points were crucial to winning the patent litigation.

Although Fleming's examination of the Edison effect was indeed essential to the invention of the thermionic valve, this was not due to some essential characters of the effect itself; it was due to the specifically Maxwellian context in which he conducted his research. That context led Fleming to construct a circuit in which an Edison lamp, a probe, a battery, and a galvanometer were uniquely combined. In the 1880s and the 1890s, this special circuit had meaning for the Maxwellian Fleming; it made little sense to others. It was *this* circuit that Fleming transformed into the thermionic valve in 1904. This explains why it was Fleming, rather than anyone else, who invented the valve, though the Edison effect was well known. The effect may have been known in the late 1880s and the 1890s, but not in the specific form in which Fleming used it.

The termination of his scientific advisorship to the Marconi Company in December 1903 compelled Fleming to change his research style significantly, and his efforts to regain credibility at the Marconi Company were crucial for the invention of the thermionic valve.

Possible uses for the thermionic valve were not obvious when it was first fabricated. Fleming actually intended it to be used to measure high-frequency alternating current in the laboratory. Marconi, not Fleming, transformed the valve into a practical receiver. Like the first transatlantic experiment and the Maskelyne affair, the thermionic valve was a contested artifact that Fleming and Marconi used to build credibility for their new wireless regime.

A New Effect in Edison's Laboratory (the "Edison Effect")

One drawback of Edison's early electric lighting system was the blackening of carbon lamps, which seriously diminished the system's efficiency. In 1879–80, Edison and his assistants in the Menlo Park laboratory noticed a thin, white shadow of the carbon filament (the "phantom shadow," they called it) on a portion of the inner surface of the bulb where no carbon was deposited. They also noticed that the shadow was more distinct near the positive side of the filament than near the negative side. It seemed that carbon was projected from the positive side of the filament in all directions. However, since their aim was less to understanding the mechanism that created the phantom shadow than to prevent blackening of the lamp, Edison and his assistants devised (in 1880) a special lamp in which an extra metallic plate was inserted to absorb the blackening carbon. From this point on, lamps with an extra metallic plate were common in the Menlo Park lab.[2]

In October 1883, one of Edison's assistants connected a galvanometer to the "anti-carbon plate" and measured the current through it. This was a natural extension of Edison's various attempts since 1880 to minimize the blackening, but it led to an unexpected result. When the added plate was connected to the positive side of an incandescent carbon filament, the galvanometer between them indicated a current of a few milliamperes. When the added plate was connected to the negative side of the carbon filament, a much smaller current flowed, even though the voltage was high enough to melt the filament. After this discovery, the phantom shadow was overshadowed by the asymmetrical current. Edison, ever attuned to practical issues, immediately thought to utilize this effect. He designed a regulator to indicate the overriding current at the end of the mains. Edison's regulator (figures 5.1, 5.2) measured the current between the filament and the anti-carbon plate, from which the fluctuation of the main's current was roughly estimated.[3]

It was widely recognized at that time that current flowed regardless of whether the metal plate was connected to the positive or to the negative terminal of the filament (figure 5.3). This is corroborated in Edison's patent specification, where, in describing the circuit connection, he specifies that the "conducting substance" (i.e., the metal plate) is "connected outside of the lamp with one terminal, *preferably the positive one,* of the incandescent conductor." Edison preferred the positive connection to the negative one because the former produced much more current. One of the earliest printed records of the Edison effect also supports my contention. Volume 38 (1884) of *Engineering,* a leading British magazine, contains a short comment on the Edison effect: "When the lamp was in action, if a galvanometer was connected between the electrode and one terminal of the filament, a current was observed which changed in direction according as the + or – terminal of the carbon was connected to the instrument. . . . The current was many times stronger when the + pole of the carbon was that connected." Edwin J. Houston, Elihu Thomson's partner in the Houston and Thomson Company,

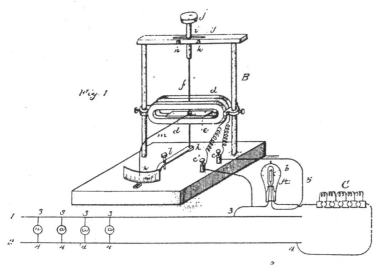

Figure 5.1
Edison's utilization of the current between the filament and the middle plate.
Source: Edison, US patent 307,031 (1884).

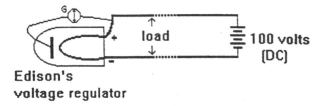

Figure 5.2
A schematic representation of figure 5.1.

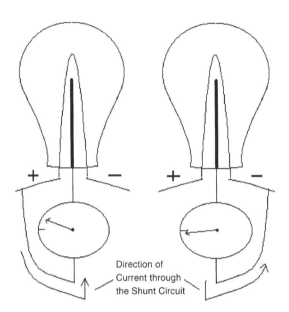

Figure 5.3
The Edison effect in the early 1880s, with a feeble reverse current in the negative connection.

spoke of the Edison effect at the 1884 meeting of the American Institute of Electrical Engineers (which had been founded the same year). From his experiments on lamps with a platinum plate, Houston said, he had noted that, even in the negative connection, "we also [had] a current flowing through the galvanometer but this time in the opposite direction," whose "amount is much less, being but about $\frac{1}{40}$ of the amount of the first circuit."[4] According to Houston, current flowed "from the place of higher to lower potential"; in such a high vacuum as existed inside Edison's lamp, however, another condition should also be considered.

In the late 1870s, William Crookes demonstrated convincingly that in high-vacuum discharge negatively charged molecules appeared to proceed from the negative to the positive electrode of a vacuum tube. This phenomenon was called "Crookes's effect" or "molecular bombardment."[5] Whether or not the phenomenon in Edison's lamp was the same as Crookes's effect was an open question; however, anyone who wanted to explain the former would have considered Crookes's effect, since there was a high vacuum in both Edison's lamp and Crookes's tube. The current in the positive connection flowed "as if it came from the platinum" (figure 5.4). This was hard to under-

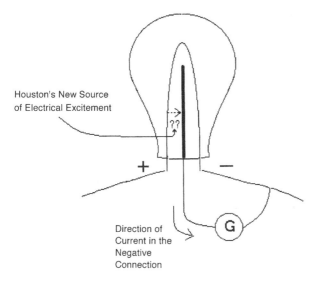

Houston's New Source
of Electrical Excitement

+   —

Direction of
Current in the
Negative
Connection

G

Figure 5.4
Houston's "new source of electrical excitement."

stand, since the carbon filament appeared to emit molecules. Though the mechanism was mysterious, Houston (1884, p. 3) suggested that "if we conceive, as I think most probable, a flow of molecules passing from the platinum to the carbon, then the phenomena may be readily explained as a Crookes's effect"—more precisely, as a kind of Crookes's effect—for the molecules proceeded from the plate charged positively. However, this left unsolved the question of the reverse current in the negative connection, in which a feeble current seemed to flow from the positive pole of the carbon filament to the plate instead of the other way round. But why did this small amount of electricity appear to jump from the filament to the plate across the vacuum, leaving the well-defined track of filament and "overcom[ing] the current from the source supplying the lamp" (ibid.)? It seemed that "in some way the molecular bombardment against the platinum plate may produce an electric current," but what caused this bombardment against the plate was uncertain. Though Houston did not offer any theory as to the origin of this current, he confessed (1884, p. 4): "For my own part I am somewhat inclined to believe that we may possibly have here a new source of electrical excitement, which might exist in the space between the plate and the filament."

Houston apparently neglected the possibility that the carbon molecules projecting from the filament were charged. "If [the current] flowed as though it went from the carbon to the platinum," he once said (Houston 1884, p. 2), "then it might be ascribed to various causes; it may be electricity flowing through empty space, which I don't think probable." William Preece, who had attended a meeting of the British Association for the Advancement of Science in Montreal in the autumn of 1884, was present at the meeting where Houston read the paper. Preece pointed out that the direction (or the sign) of currents was nothing but a convention. Quoting some recent experiments of Crookes and William Spottiswoode,[6] he concluded that "if electricity does flow at all in any direction, that direction is rather from the lower to the higher potential" (Preece 1884, p. 6). Even though Preece did not specify the mechanism of the effect on which Houston reported, his idea implied that the relatively large amount of current in the positive connection was simply the flow of negative electricity from the minus pole of the carbon filament to the platinum plate, rather than the other way round. This explanation fit well with the observed phenomenon—that is, the carbon deposit on the glass of the bulb. Also, the current in the positive connection was explained by the ordinary Crookes's effect—that is, the molecular bombardment from the negative to the positive electrode in a high vacuum.

But how did Preece account for the feeble reverse current in the negative connection? He knew this to be troublesome: "When we have platinum, it is a very difficult thing to prove that there exists Crookes's effect." (ibid.) To be consistent, Preece should have argued that negative electricity had to proceed from the (–) platinum plate to the (+) carbon filament across the vacuum. But what caused this jump across the high-resistance vacuum? Since Preece believed that "to produce a current we must have two points separated from each other by matter," he asked "What is that matter, and where is that matter?" (ibid.) Preece mentioned Lodge's finding that conductivity inside a vacuum tube was greatly increased by producing an electric spark near it, thereby creating a fog—a matter—inside it. According to Preece, increasing conductivity by means of a spark enabled Crookes's effect to take place in a high vacuum. If this spark-like "new source of electricity" existed in Edison's lamp, Preece conjectured, it could explain the strange effect. In any event, Preece (1884, p. 7) seemed to believe that the currents in both the positive and the negative connections could be explained by the Crookes's effect.[7]

Preece visited Edison's Pearl Street Station in New York and obtained several lamps that were specially constructed to examine the strange effect.

Preece brought these to England. After experimenting with them for several months, he reported his results to the Royal Society in March 1885. It was in this report that Preece coined the term "Edison effect."[8] The most original aspect of this paper was Preece's careful measurement of the current in the galvanometer connected to the plate and the positive pole of the filament. Preece also measured the resistance of the space between the plate and the negative pole of the filament. From the values of the current and the resistance, he estimated the apparent voltage applied across the space. He compared this voltage with the voltage of the main batteries that supplied current through the filament. The results did not reveal anything particularly interesting to Preece. In one lamp, the seemingly applied voltages across the vacuum were relatively stable, around 10 volts, despite the increase of the main voltage from 44 to 80 volts. But when the latter reached a critical point, the former increased dramatically to 95 volts, almost the same as the latter (95.49 volts). In other cases, the voltage across the vacuum remained far below the main voltage, and it decreased when the latter increased. In other cases, the measurement of the shunt current was too unstable to be used for any estimations.[9]

Concerning the theoretical side of the Edison effect, Preece remained as convinced as before. Preece observed current flow in both connections, although one of Edison's assistants had told him that "no shunt current or a very slight one could be observed" in the negative connection. Only one lamp did not show any current in the negative connection, but even in this case Preece (Preece 1885, p. 222) remarked "Doubtless I should have got both effects if I could have raised the emf." However, he did not suggest any mechanism to explain the negative current; he simply ignored it. He did explain the positive current, which he thought was due to the projection of carbon molecules from the negative side of the filament to the plate— that is, Crookes's molecular bombardment. He supported his view by showing, with the use of several lamps, that no shunt current was observed when the (supposed) carbon molecules were prevented from reaching the plate (Preece 1885, pp. 230).

Fleming's Early Research on the Edison Effect

Fleming, who had worked as consulting electrician to the British Edison Company since 1882, also noticed the strange "molecular shadow" in Edison lamps. In a short paper read before the Physical Society in 1883 he addressed the curious shadow of a filament inside the carbon-blackened lamps. In a second paper (1885a), he stated that the shadow was created by the molecular projection of either the carbon or the copper that was used to make the clamp holding the carbon filament. He also suggested an application for the carbon deposit: the production of a thin metallic film for optical experiments. However, Fleming's first two papers did not contain any deeper theoretical insight into the phenomena. Fleming neither constructed nor used a special lamp with an extra plate in his work. Therefore, unlike Preece, Fleming did not measure the current in the vacuum. In this sense, Fleming did not really experiment on the Edison effect; he only observed and speculated on the phantom shadow.[10]

Why, then, did Fleming continue experimenting with the Edison lamps and eventually surpass the work of other scientists and engineers on the Edison effect? One reason is that, for Fleming but not for others, the Edison effect posed new theoretical and practical challenges. William Preece believed that he had explained the Edison effect in theory: it was simply due to the Crookes's effect. Further, he found no practical application for it, because, in his words (Preece 1885, p. 229), "the current is weak and variable, and it is scarcely reliable enough to be useful for practical purposes as was hoped by its discoverer." Any scientist interested in the nature of conduction in a vacuum would have used Crookes's lamp (which used a metallic plate as a cathode), not the carbon-filament Edison lamps with their additional plates. Edison lamps could be constructed only in the workshop of a specialized lamp factory, and not in the more general-purpose laboratory of a physicist or an engineer.[11]

From the beginning, Fleming perceived a difference between the molecular shadow in Edison lamps and the Crookes effect: in the former, "we are dealing not with an induction-coil discharge, but with a comparatively low potential" (Fleming 1883, p. 284). In modern terminology, Crookes's effect is a cathode discharge, whereas the Edison effect is a thermionic emission, although the difference between the two was barely recognized in the 1880s. Besides this difference, Fleming found the notion that the current involved the motion of molecules particularly interesting, because it was not a basic Maxwellian assumption. (Maxwellians conceived current as, in its essence, changing electric displacement.) This is why Fleming emphasized research on the "constituents of the current" as a primary topic when he proposed the establishment of a National Standardizing Laboratory in 1885.[12]

Fleming also had a material advantage. After he became professor of Electrical Technology at University College London, in 1885, he retained his position as scientific advisor to the Edison-Swan Company, which specialized in the construction of incandescent lamps. As scientific advisor to the company, Fleming's main task was to construct photometric standard lamps, which did not exhibit visible carbon deposits for a long time. Above all, he could easily have special lamps constructed in the factory of the Edison-Swan Company in Newcastle-upon-Tyne and, later, at Ponders End (Fleming 1883, p. 284).[13]

In 1889, Fleming did some research for the Edison-Swan Company on the specific conductivity of carbon filaments. In the course of that work, he performed a series of experiments on the Edison effect. He read a paper on that topic before the Royal Society, and he gave a public lecture at the Royal Institution (Fleming 1890a,b). In his papers, Fleming presented some novel results and conclusions. First, he eliminated the feeble (and mysterious) current in negative connection, for which neither Preece nor anyone else could account. As we have seen, Houston assumed a new kind of electricity to explain the current in the negative connection, while Preece ignored it. This made the Edison effect theoretically unstable. Yet this new source of electricity became unnecessary when Fleming (1890a, pp. 119, 120; 1890b, p. 38) showed that "if the metallic plate is connected with the positive electrode of the lamp, the galvanometer indicates a current of several milliamperes, while it is connected with the negative electrode of the lamp, it shows *no sensible* current."[14] Fleming reported that when the lamp had a good vacuum the current between the plate and the negative electrode was not greater than 0.0001 milliampere. When the vacuum was poor, current was again detected even in the negative connection. Fleming had consequently made the incomprehensible current in the negative connection into an artifact of poor instrumental design. Though it remained difficult to understand why current flowed in the negative connection, this did not matter much, first because conduction in poor vacuum remained a complex and contentious process, both empirically and theoretically, and second because Fleming could make this feeble negative current vanish with a better instrument. The pathological current in the negative connection became a secondary discharge process occurring in a poor vacuum, which had nothing to do with the current in the positive connection. Fleming's stabilization of the Edison effect was, then, due largely to the high quality of the vacuum lamps that were available to him. And these excellent lamps were available to him because he was a scientific advisor to the Edison-Swan Company and a close friend of Charles H. Gimingham, an electrician at the Edison-Swan factory, who had constructed a special vacuum pump for the purpose of making better incandescent lamps.[15]

On the basis of experiments in which various parts of the filament had been shaded, Fleming decided that the Edison effect was due to the projection of negatively charged carbon molecules (which he called "molecular electrovection") mainly from the "negative" leg of the filament to the metal plate. These negatively charged carbon particles, which permeated the lamp, reduced the potential of the inserted plate to the same level as that of the negative leg of the filament. A current (whose conventional direction is the direction of positively charged electricity) thus flowed in the shunt circuit only when the plate was positively charged, thereby discharging the negative carbon molecules that had hit it (Fleming 1890a, p. 122; Fleming 1890b, p. 43). Fleming not only identified the carriers of electricity with ionized carbon particles; he also explained why carbon could retain only a negative charge by resorting to an experiment performed by Frederick Guthrie, his former mentor at the South Kensington Science School. Guthrie had shown that an iron ball, when heated, retained only negative charges (Fleming 1890b, p. 46).[16] By analogy to Guthrie's small hot iron balls, Fleming argued that hot carbon particles could retain only negative charges. This view was entirely consistent with the Maxwellian understanding of electric charge: just as the negative charge on a hot iron ball did not necessarily mean that electric charge existed as a substance in its own right, so the negative charge on hot carbon particles did not mean that they bore independent charge carriers. The charge on the carbon particles could still be seen as having been created as an end effect of fields on them.

Conceptualizing and Utilizing "Unilateral Conductivity"

During his research on conductivity in the Edison lamp, Fleming paid increasing attention to a curious phenomenon in the "half inch of highly vacuous space between the hot carbon conductor and the middle plate." He reasoned that it was caused by negatively charged carbon particles emitted from the filament when they fell on (and thereby discharged) the positively charged plate. Fleming tried to test this by connecting a small condenser to the shunt circuit. When the positive side of the condenser was connected to the middle plate, it turned out, as he expected, that the condenser was discharged at once; when the connection was reversed, discharge never took place. (See figure 5.5.) This "led [him] to examine the condition of the vacuous space between the middle metal plate and the negative leg of the carbon loop," and in the context of Maxwellian electromagnetic theory the nature of the conductivity of this space was of utmost importance. To probe it, two things were necessary: a closed circuit including the vacuum and a source of emf. The circuit Fleming used to study conductivity is illustrated in figure 5.6. A Clark cell (Ck), the middle plate (M), the space, and the negative leg of the filament constituted a closed circuit. A large external emf rendered the filament itself incandescent. Current was detected in the galvanometer when the positive pole of the Clark cell was connected to the plate, but when the connection was reversed no current was detected. The intervening space apparently possessed "a kind of *unilateral conductivity*, in that it will allow the current from a single galvanic cell to pass one way but not the other." The space behaved neither like normal circuit elements nor like a source of electricity. It seemed that only negative electricity could be sent from the filament to the plate.[17]

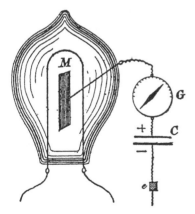

Figure 5.5
Fleming's experiment to show that the plate *M* was negatively charged. Source: Fleming 1890a, p. 43.

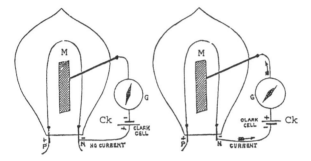

Figure 5.6
Fleming's experiment to probe the conductivity between plate M and the filament.
Current flowed in the galvanometer (G) only when the positive pole of the Clark
cell (Ck) was connected to M, thereby revealing the space's unilateral conductivity. Source: Fleming 1890a, p. 44.

Fleming located his observation of unilateral conductivity in the capacity of negatively charged carbon molecules to flow through space from the
(hot) filament to the (cold) plate. The relationship between the filament
and the plate was thus asymmetrical: the filament was heated by the main
current, while the plate was not, because the current there only discharged
the negative electricity accumulated on it; the filament was inserted into a
closed circuit as an ordinary circuital element, while the plate formed part
of a closed circuit only through a connection across the space. Focusing
on the shunt circuit proper, which consisted of the plate, the filament, the
space between them, and the Clark cell, Fleming thought that the conductivity of the intervening space had increased to such an enormous extent
that even a single cell could flow current through the space in a specific
direction.

    This phenomenon particularly interested Fleming, since as an unknown,
yet ambitious, science student in the 1870s, he had been interested in the
production of induced currents in liquids. He received modest grants for
this research and read short papers before the British Association and the
Royal Society, but when he submitted his full paper to the society's
Philosophical Transactions for publication one of the referees (James Clerk
Maxwell) recommended that he perform more experiments on the production of induced currents in gases. Fleming, who had resigned his science
mastership at Cheltenham College to go to Cambridge University, had neither the time nor the laboratory space to pursue this matter. In addition, as

Maxwell mentioned, it was very difficult to induce a current in a gaseous medium, owing to the latter's high resistance. Since that time, the question of how to reduce the resistance in a gaseous medium had been on Fleming's mind.[18] Conductivity change in a gaseous medium would have been important to Fleming in view of Maxwell's remarks to him. It is therefore not surprising that the final question he asked concerned the relation between unilateral conductivity (which was intrinsically puzzling to many Maxwellians[19]) and increased conductivity. What would happen to the conductivity of the space if the middle plate were itself heated and rendered incandescent? Would unilateral conductivity persist? One way to make the middle plate incandesce might be to concentrate a powerful beam of light on it (the photoelectric effect).[20] Fleming, however, took the easier path of substituting another carbon filament for the platinum plate. This abolished the inherent asymmetry of the lamp's previous configuration (figure 5.7).

Compare figure 5.7 and figure 5.6. The two circuits within the lamp differed only in that the voltage across the small filament M was lower than across the main filament C. Current was now detected in the galvanometer as long as both C and M were incandescent, whether the galvanometer was connected to + or to –. Unilateral conductivity had now vanished, while the increase in conductivity survived. As can be seen in figure 5.8, when voltage B_1 is equal to B_2, and when both filaments are incandescent, the space V becomes such a good conductor that a Clark cell of about 1.5 volt can flow current through it (Fleming 1890b, pp. 44–45).

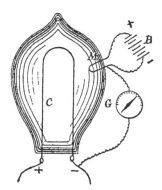

Figure 5.7
Fleming's experiment with an incandescent middle phase (M). Source: Fleming 1890a, p. 45.

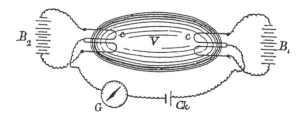

Figure 5.8
Fleming's demonstration technology for showing that the space V between the two filaments had such a low resistance that a Clark cell (Ck) could produce current across it. Note that the asymmetry in figure 5.6 has vanished. Source: Fleming 1890a, p. 44.

Fleming had come full circle. He had stabilized the Edison effect by eliminating the incomprehensible current in negative connection and by creating a unilateral condition between a middle plate and a filament. Securing good vacuum lamps was essential to this success. However, Fleming found a physical asymmetry between the plate and the filament, which seemed to be intimately connected to the increase in conductivity there. By eliminating the asymmetry, he emphasized large conductivity and downplayed unilaterality. His demonstration apparatus (figure 5.8) accordingly embodied Fleming's special concern with the increase in conductivity in the gaseous medium. However, as we can easily see, there was a different route of artifactual development by utilizing unilaterality. This was what led Fleming to the thermionic valve in 1904, but in a different context. It was not until around 1900, when Fleming adopted the electron theory and thereby discarded his Maxwellian commitment, that he could transform this unilaterality into the valve.

The Canonical Story about the Valve and Its Problems

Fleming applied for a patent for the thermionic valve in 1904. Lee de Forest applied for a patent for the audion (disregarded by Fleming as a mere insertion of one more electrode in his own valve) in 1906. Neither device was widely used until the early 1910s. In the 1910s, however, the audion was employed as a continuous-wave generator as well as an amplifier, and this deeply affected the entire field of radio engineering. To monopolize or to have a major share in the patent on the vacuum tube had by then become a matter of life and death for many companies. Fleming first reminisced about his own invention during the resulting patent litigation between the Marconi and de Forest companies. The following quotation, which became the canonical history of the thermionic valve's origin, comes from Fleming's 1920 Friday Lecture at the Royal Institution (Fleming 1920, pp. 167–168). Fleming began with the deficiencies of the detectors that existed around 1904:

Before 1904 only three kinds of detectors were in practical use in wireless telegraphy—viz. the coherer, or metallic filings detector, the magnetic-wire detector, and the electrolytic detector. The coherer and the electrolytic detectors were both rather troublesome to work with on account of the frequent adjustments required. The magnetic detector was far more satisfactory, and in the form given to it by Marconi is still used. It is not, however, very sensitive, and it requires attention at frequent intervals to wind up the clockwork which derives the moving iron wire band. In or about 1904 many wireless telegraphists were seeking for new and improved detectors.

He then spoke of his efforts to find a new detector, which eventually led him to the construction of the valve:

I was anxious to find one which, while more sensitive and less capricious than the coherer, could be used to record the signals by optical means, and also for a personal reason I wished to find one which would appeal to the eye and not the ear only through the telephone. Our electrical instruments for detecting feeble direct or unidirectional currents are vastly more sensitive than any we have for detecting alternating currents. Hence it seemed to me that we should gain a great advantage if we could convert the feeble alternating currents in a wireless aerial into unidirectional currents which could then affect a mirror galvanometer, or the more sensitive Einthoven galvanometer. There were already in existence appliances for effecting this conversion when the alternations or frequency was low—namely, one hundred, or a few hundred per second.

For example, if a plate of aluminum and one of carbon are placed in a solution of sodic phosphate, this electrolytic cell [the Nodon rectifier] permits positive electricity to flow through it from the aluminum to the carbon, but not in the opposite direction. . . . But such electrolytic rectifiers, as they are called, are not effective for high frequency current, because the chemical actions on which the rectification depends take time. After trying numerous devices my old experiments on the Edison effect came to mind, and the question arose whether a lamp with incandescent filament and metal collecting plate would not provide what was required even for extra high frequency currents, in virtue of the fact that the thermionic emission would discharge the collecting plate instantly when positively electrified, but not when negatively. . . . I found to my delight that my anticipations were correct, and that electric oscillations created in the second coil by induction from the first were rectified or converted into unidirectional gushes of electricity which acted upon and deflected the galvanometer. I therefore named such a lamp with collecting metal plate used for the above purpose, an oscillation valve, because it acts towards electric currents as a valve in a water-pipe acts towards a current of water.

After 1920, Fleming repeated this recollection in books, articles, and lectures, sometimes with variations. At times he dramatized the moment when the "happy thought" (i.e., the thought of utilizing the unilateral conductivity for high-frequency oscillations) struck his mind. At other times, he stated that he had tried to use an extremely delicate siphon recorder with a Morse printer, which had been used as a detector in submarine telegraphy; since the siphon recorder required direct current, he needed to devise something to convert a high-frequency alternating current into a direct one.[21] Nevertheless, the fundamental structure of his story remained unchanged, always involving the following three stages: the trouble with existing detectors and the consequent need for a new one around 1904, his recognition of "rectification" as a new method of detecting oscillations, and how he had found a method of rectification in his prior research on the Edison effect in bulbs.

Fleming's three stages support his claim to have invented the thermionic valve. The story is so neatly and convincingly laid out that few historians have doubted its accuracy. In regard to the demand for stable detectors around 1904, George Shiers (1969, p. 109) argues that Marconi's detectors "were not satisfactory for regular and dependable service; a new and better detecting device, or signal rectifier, was urgently needed." Concerning the importance of Fleming's prior research on the Edison effect, Hugh Aitken (1985, p. 205) asserts that "Fleming's valve was a linear descendent of a device that had no connection with wireless telegraphy at all; this was the famous Edison Effect." Though Fleming emphasized the scientific nature of his research on the Edison effect, the same point was sometimes used to devalue his originality. Gerald Tyne (1977, pp. 40–51) commented that "what [Fleming] did was to apply the Edison-effect lamp, patented by Edison in 1884, to the rectification of high-frequency oscillation."

Three points in Fleming's recollection are disputable.

First, it is not entirely clear that there was a compelling demand for a new detector around 1904 . Marconi's magnetic detectors were more stable than coherers, and they had already proved especially useful for long-distance wireless telegraphy. As Fleming himself mentioned, electrolytic detectors were popular in Germany and in the United States. They could be used not only with a telephone but also with a galvanometer. The Lodge-Muirhead wireless system employed a very delicate siphon recorder, devised by Alexander Muirhead, and printed signals on a Morse tape.[22] In fact, for that reason the thermionic valve, when first used as a detector at the Poldhu station in the summer of 1905, was used only in special cases, such as when someone wanted to reduce the influence of atmospheric electricity.

Second, Fleming's research during 1903–04 apparently had little to do with detectors in wireless telegraphy. At that time, he was interested mainly in high-frequency measurements—measurements of the inductance capacitance, resistance, current, frequency, wavelength, and the number of sparks. It will be argued in the next section that Fleming's valve resulted partly from his concern with the measurement of feeble, high-frequency alternating current.[23] Such a concern was, in one sense, connected to the detection of waves, since a precise instrument for current measurement could be used as a detector. But a meter is designed to produce a number, whereas a detector is designed to detect pulses. Metrical character and easy calibration were essential to the former; sensitivity and stability for use in the field were important to the latter.[24]

Third, Fleming's recollection, like many other stories of the thermionic valve, does not consider the specific context in which his research was performed. In December 1903, Fleming's scientific advisorship to Marconi was terminated against Fleming's wishes. As we saw in chapters 3 and 4, there had been tension between Fleming and Marconi over the credit for the first transatlantic telegraphy for some time, and the Maskelyne affair in June 1903 hurt Fleming's credibility seriously. Fleming's dual role as Marconi's scientific consultant and an authoritative witness of Marconi's secret demonstration could no longer be sustained. Fleming, who wanted to build credibility as an "ether engineer," desperately wished to recover his connection to the Marconi Company. His efforts to this end help explain a notable change that took place in his engineering style at that time. Before 1903, Fleming's style was distanced from the inventing or patenting of devices, but in 1904 he patented two inventions: the "cymometer" (a wave-measuring instrument) and the thermionic valve. As we will see, these two devices (particularly the valve) enabled Fleming to resume his connection with the Marconi Company.

Scientific and Corporate Contexts, 1903–04

In 1896, at the Cavendish Laboratory, Ernest Rutherford experimented on the effect that electromagnetic waves magnetized a piece of demagnetized iron and demagnetized a piece of magnetized iron. In 1902, on the basis of

Rutherford's finding, Marconi invented a practical magnetic detector with a telephone as a signal indicator. Ordinary DC galvanometers could not detect alternating signals; they were too slow to react to the magnetizing action before it was annulled by the demagnetizing one. In December 1902, pursuing a rather different goal than Marconi, Fleming and his assistant Arthur Blok decided to investigate how the Rutherford effect might be made to work with a galvanometer.[25] Suppose, they thought, electromagnetic waves were allowed only to demagnetize the iron, which was then magnetized again by some other means. The effect of electromagnetic waves on the magnetic detector could then be registered by an ordinary DC galvanometer. In other words, if electromagnetic waves acted in only one way (either to magnetize or to demagnetize), this effect could be converted into electrical signals to be detected with a DC galvanometer. After much trial and error, Fleming and Blok constructed a workable device, which Fleming described before the Royal Society in March 1903 (figure 5.9). The instrument used

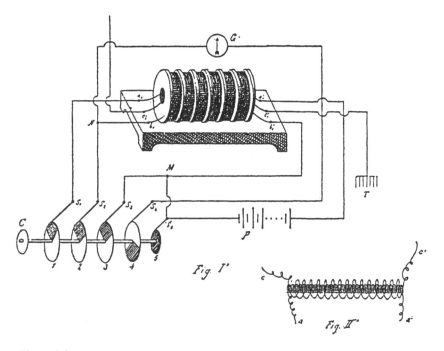

Figure 5.9
Fleming's quantitative magnet meter for measuring high-frequency alternating current. Source: Buscemi 1905.

wire bundles as a magnetizing (and demagnetizing) core, and it had a set of complex commutators that controlled the one-way magnetic action. The sophisticated motion of the commutator disks endowed the device with the ability to *"produce the effect of a continuous current in the galvanometer circuit,* resulting in a steady deflection, which is proportional to the demagnetizing force being applied to the iron, other things remaining equal" (Fleming 1903b, p. 400).

Fleming listed several major uses for the device. Primary among them was "comparing together the wave-making power of different radiators." The power of the wave generated depended, for example, on the shape and materials of the ball discharger. It also depended on the rotation of the balls, and on the dielectrics (such as oil) placed between them. In order to determine under which conditions the radiator would produce the best results, an instrument that could measure the effect in numerical terms was badly needed. In his March 1903 Cantor Lecture at the Society of Arts, for example, Fleming mentioned the need for a device that could measure the wave energy. "It is only by the possession of such an instrument" he said, "that we can hope to study properly the sending powers of various transmitters, or the efficiency of different forms of aerial, or devices by which the wave is produced."[26]

Used as described in the preceding paragraph, Fleming's new device was not a detector but rather as a kind of high-frequency AC galvanometer—a meter. As a detector, it proved cumbersome and less efficient than other detectors of the day. Nor did Fleming produce the device in the first place to serve as a detector; he always had metering in mind. In any case, the instrument was not widely used. Outside Fleming's own engineering laboratory at University College, it was used only by an Italian scientist, V. Buscemi, in examining absorptions of electromagnetic waves in various dielectrics. The reason for this neglect seems to be the instrument's lack of stability. It turned out to be very difficult to calibrate because of complex commutator actions (Buscemi 1905).[27]

In 1903, Fleming redirected his research toward devising instruments that could directly measure the effects of high-frequency currents. In theory there was no reason why alternating current meters in power engineering could not be used for that purpose. Two different types of AC meters were used in power engineering: the dynamometer and the hot-wire ammeter. Each had drawbacks. Currents below 10 milliamperes (encountered often in high-frequency measurements) could not be measured with ordinary AC dynamometers.[28] Fleming therefore tried to design an AC ammeter in which fine wires would sag when heated by high-frequency currents. His hot-wire device could measure 10 milliamperes within an accuracy of 2–3 percent, and could detect 5 milliamperes. It could be calibrated easily by means of the potentiometer method, a standard method used to calibrate ordinary AC galvanometers. But it could not measure below 5 milliamperes, and this was a serious defect. Shortly after Fleming developed the hot-wire ammeter, the British engineer William Duddell, well known for his ingenuity in designing scientific instruments, exhibited his thermo-galvanometer, which employed a tiny thermoelectric junction and which, Duddell (1904) argued, could measure even 0.1 milliampere. Fleming (1904a) countered that Duddell's device was hard to calibrate, since it was extremely difficult to keep the thermo-junction and its thermo-forces constant.

In December 1903, when Fleming's first three-year advisorship expired, the Marconi Company did not renew his contract. In subsequent communications with Marconi, Fleming emphasized the value of his research on the rotating-ball discharger in regard to power signaling, thinking Marconi would be interested. However, Marconi had already decided that it was impractical. By 1904, Fleming's relationship with Marconi and the company had virtually ended. Maintaining a connection to Marconi was vital to Fleming not merely because Fleming's advisorship gave him money and recognition but also because his department and laboratory at University College were then becoming the first such facilities in Britain to be dedicated to radio engineering. Without a link to Marconi, Fleming's laboratory would not fully develop into a research laboratory for radio engineering.[29] Fleming did not know how to revive the connection, but he felt that useful inventions for wireless telegraphy might help.

These circumstances altered Fleming's style. As I remarked above, before 1903 Fleming had not patented a product of his own research. Even when he significantly improved the design of existing potentiometers (in 1884), or when he discovered unipolar induction (in 1891), he was not enthusiastic about patenting devices or in using his discoveries for practical ends. Fleming's strength was in explaining technical problems in scientific and mathematical language, and he had been satisfied with his fame as a mediator between science and engineering. By 1904, however, he felt heavy pressure to invent something useful in order to restore his links to Marconi.

This pressure forced him to try transforming some of his laboratory devices into technological appliances. He rapidly succeeded with two artifacts. The cymometer, a wavelength-measuring device, had grown out of a resonance effect on a long coil, known as Seibt's coil, that had been used as a demonstration device to show the peaks and valleys of the electromagnetic waves induced on it. Fleming transformed Seibt's coil into a practical and portable wavelength-measuring device.[30] The second successful device was the thermionic valve, which will be examined in the following section.

Rectifiers, Photometry, and the "Happy Thought"

The rectifiers of the early 1900s could be classified into two groups: those for "heavy" low-frequency currents in power engineering, and those for feeble high-frequency currents in wireless telegraphy. Converting alternating into direct current had been an important goal in power engineering, and several kinds of rectifiers (including rotary converters, mercury lamps, vacuum tubes, and electrolytic cells) had been invented for that purpose.[31] On the other hand, no efficient rectifiers existed for wireless telegraphy. Oliver Lodge and Alexander Muirhead sometimes employed Lodge's mercury lamp rectifiers in wireless receivers to measure the effect of transmitters. However, the mercury lamp rectifier, originally designed to collect dust in the air, proved to be inefficient for wireless telegraphy (Lodge 1903).[32]

Though not invented for rectifying purposes, electrolytic detectors could be used to rectify high-frequency oscillations. The mechanism that governed electrolytic detectors was much debated in the early 1900s, and the argument that the electrolytic breakdown of anode polarization by incoming electromagnetic waves produced a rectifying effect competed with the claim that a thermal process was responsible. The German scientist M. Dieckmann argued in 1904 that the detector was actuated not by thermal process but by electrolytic depolarization (Dieckmann 1904). V. Rothmund and A. Lessing, two other Germans who were experimenting on the same subject, obtained the same conclusion and published their results in the September issue of *Wiedemann's Annalen* (Rothmund and Lessing 1904). Meanwhile, Dieckmann's article was translated into English and printed in *The Electrician*.[33] Rothmund and Lessing, like Dieckmann, employed an ordinary DC galvanometer in their experiments to demonstrate high-frequency rectification by electrolytic detectors. Since Fleming had previously sought ways to create the effect of continuous current from high-frequency oscillations in order to measure wave power, these German works may have attracted his attention. Yet a more direct motivation, it seems, was a paper by Albert Nodon, a French electrician, titled "Electrolytic Rectifier: An Experimental Research," which was read at the St. Louis International Electrical Congress in the summer of 1904 and published in the October 14, 1904 issue of *The Electrician* (Nodon 1904). Nodon described the characteristics and the efficiency of some aluminum cell rectifiers he had designed for use in power engineering. Nodon's rectifiers consisted of two metals (one usually aluminum) dipped in an electrolyte (carbonate of ammonium, for example). With low-frequency (42–82 Hz) alternating current, Nodon obtained 65–75 percent efficiency in rectification. In his article, Nodon described his rectifier as an electrolytic "valve" (which suggests the origin of Fleming's name for his own device) and used such terms as "valve effect" and "action of the valve." Fleming tried to use Nodon's valve to replicate the rectifying effect with high-frequency oscillations, but after many experiments he concluded that it was ineffective at high frequencies.[34]

The case of the Nodon valve illuminates why Fleming later said repeatedly that electrolytic cells or rectifiers were inefficient for high-frequency rectification. Lee de Forest once criticized Fleming for this, because most electrolytic detectors in wireless telegraphy could in fact do the job. However, by "electrolytic rectifiers" Fleming specifically meant Nodon's valves of the sort used in power engineering, not the electrolytic detectors used in wireless telegraphy. Furthermore, Fleming's attention to the Nodon valve supports my claim that between 1903 and 1904 he was less concerned with new detectors than with finding an instrument for measuring high-frequency currents. If he had been interested in primarily detectors, he would have started his experiments with ordinary electrolytic detectors of the sort used in wireless telegraphy, rather than with Nodon rectifiers of the sort used in power engineering. (The latter had nothing to do with wireless telegraphy.) And although Fleming once mentioned that he had invented the thermionic valve in October 1904, in fact he did not learn of the Nodon rectifier until the middle of that month; this suggests that his discovery may have occurred somewhat later. A short letter to *The Electrician* written in 1906 by Fleming himself supports a later date: "In November 1904 [I] discovered that this unilateral conductivity [in the lamp] held good for high frequency current, which is not the case for electrolytic rectifiers."[35]

After his failure with the Nodon valve, Fleming pondered other ways to rectify high frequencies. A "happy thought" took him back to the unilaterally conducting space inside the lamp. In the 1890s, as we have seen, Fleming had tried to eliminate unilaterality. In 1904, he could utilize it for rectification. Only now returning to his earlier work (the development of which is illustrated by figures 5.6–5.8), Fleming produced the circuit illustrated in figure 5.10, having asked his assistant G. B. Dyke to take "out of a cupboard one of [his] old experimental bulbs." The circuit immediately proved workable.[36]

Why, at this precise moment, did Fleming conceive of using the lamp for rectification—an idea that had never occurred to him in the past 20 years? Although it is scarcely possible to find a single reason for this kind of "happy thought," several factors that held only in 1904 may well be responsible.[37] We have already seen that by 1904 Fleming felt pressed by an "invention imperative." We have examined Fleming's efforts to find a suitable current meter that could rectify high-frequency oscillations. We have seen that various works on AC rectification became available to him since the summer of 1904, and one of them (the paper on Nodon's valve) may have led Fleming to the term "valve" and even to the notion of valve-like action.

The first of two further factors that are pertinent is the electron theory of Joseph Larmor and J. J. Thomson, which Fleming adopted around 1900.[38] According to the electron theory, the charge carriers in the conduction current were electrons. The Edison effect would then be interpreted as a movement of a portion of the negatively charged electrons in the vacuous space, from the negative pole of the filament to the plate. This could have changed the meaning of several of the circuit elements shown in figure 5.6. In 1890, before the electron theory, the Clark cell (Ck) functioned as a "sucker" or a discharger of negative charges on the plate (M), which accumulated on M as a result of bombardment by charged carbon particles. The galvanometer (G) measured the rate of this discharge. In 1904, in the context of electron theory, the Clark cell became a source of charge carriers, which pushed its electrons to the negative pole of the filament to propel them through the space, and the galvanometer measured the effect of electrons being driven from the Clark cell. Unilateral conductivity then became a possible means

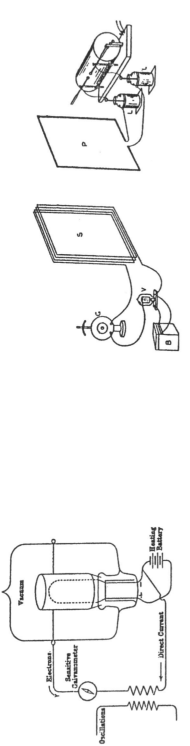

Figure 5.10
Fleming's valve in 1904, or how to use unilateral conductivity. In the right drawing, the valve (V) is connected to the galvanometer (G) to detect electromagnetic radiation. Sources: (left) *The Electrician* 54 (March 31, 1905), p. 907; (right) Fleming 1906c, p. 179.

of rectification of an AC source, if it replaced the Clark cell. But there are two problems in connecting electron theory with Fleming's thermionic valve:

• Fleming's patent specification for the valve does not mention electron theory in detail, nor does his first paper on the valve. The paper only mentions "electrons or negative ions" in a passing way (Fleming 1905, p. 487). Fleming does not quote Owen Richardson's experiments on thermionic emissions,[39] nor is there any evidence that he was aware of Richardson's research, although there is a high probability that he had heard about it.[40]
• Recent historical and philosophical discussions on scientific practice agree that "high theory" (such as electron theory in the present context) often does not affect the everyday practice of a scientist directly, because a scientist usually uses much "lower" theories (e.g., theories grounded in instruments).[41]

There is, however, something outside electron theory proper that might have stimulated Fleming, although in the absence of corroborating evidence this must remain only a suggestion. In the October 1904 issue of *Philosophical Magazine*, the aged Kelvin (1904a,b), basing his argument on the claim that the perfect vacuum was a perfect conductor, stated: "The insulation of electricity in vacuum is to be explained, not by any resistance of vacant space or of ether but by a resistance of glass or metal or other solid or liquid against the extraction of electrons from it, or against the tearing away of electrified fragments of its own substance." If Fleming had seen these papers by Kelvin (as is highly probable—since Kelvin 1904a is mentioned in Fleming 1906b), they certainly could have reminded him of his own much earlier research on the Edison effect, thereby leading him to reconceptualize unilateral conductivity in the vacuum in Edison's lamp in terms of the movement of electrons extracted from the carbon filament.

The second factor is Fleming's involvement in photometric measurement. Although Fleming's last experiment on the Edison effect was performed in 1895, lamps had always been an important material resource in Fleming's laboratory because of the photometric research being done there. As Arthur Blok recalled (1954, p. 127):

[Fleming in the early 1900s] had initiated much work on the changes in luminous emission from carbon filament lamps, whether caused by changes in the filaments themselves or by blackening of the bulbs. At the Pender Laboratory at Gower Street we were all in this and would often emerge from the photometric gallery after hours of close work with the Lummer-Brodhum blinking like owls in the noon-day sun.[42]

Preventing carbon deposition on the glass was extremely important in the construction of the standard photometric lamp. In August 1904, at a meeting of the British Association for the Advancement of Science, Fleming read an article on the standard lamp, the large size of which "prevents any sensible deposit of carbon upon it" (Fleming 1904b). This shows that the carbon deposit inside a bulb, which had stimulated Fleming's long research on the Edison effect, was still of interest to him at the moment when he constructed the thermionic valve.

Constructing the Use of the Valve

After his first successful experiment with the thermionic valve, Fleming filed a provisional specification for it (British patent 24,850, November 16, 1904). The patent was titled Improvements in Instruments for Detecting and Measuring Alternating Electric Currents. We can confirm the purpose of his invention in this provisional specification. Nowhere did Fleming mention troubles with existing detectors, nor did he allude to a putative demand for new ones. After mentioning his previous research on the measurement of high-frequency alternating current, he explained the object of the invention: "to provide a means by which an ordinary galvanometer can be used to detect and measure alternating electric currents and especially high frequency currents commonly known as electric oscillation." He then mentioned two uses for the valve: (1) "The device is especially applicable to the detection and measurement by an ordinary galvanometer of high frequency current or electric oscillations, where any form of mechanical or electrolytic rectifier [such as Nodon's valve] is useless." (2) The device "and a galvanometer may be used as a receiving instrument in wireless telegraphy." The first objective emphasized the device's use, with an ordinary galvanometer, to measure currents in the laboratory; the second suggested its possible use as a receiver in wireless telegraphy.[43]

From December 1904 to January 1905, Fleming experimented with the only valve that was working satisfactorily in order to determine its conductivity with different electrode voltages. In these experiments Fleming found that the relationship between the current and the voltage "was, to a very large extent, independent of the electromotive force creating it, and is at no stage proportional to it."[44] There was no linearity, nor was there any regularity in the characteristic curve that represents the relationship between the current (*I*) and the voltage (*V*). Rectifying power was therefore scarcely predictable, which undermined Fleming's initial hope that he had "produced" a metering device. What about the valve as a detector, then? Fleming was well aware of the valve's merits and faults as a detector; this is evident from his statement that "the arrangement [a valve with a galvanometer], although not as sensitive as a coherer or magnetic detector, is much more simple to use." Manageability was a merit; poor sensitivity was a weakness. But another merit—the ability to detect "a change in the wave-making power or uniformity of operation of transmitting arrangement"—would make up for this disadvantage. The valve, used with a galvanometer, exhibited a modest metrical character, which the coherer or the magnetic detector lacked; however, this was not sufficient to "make it useful as a strictly metrical device for measuring electric oscillations" (Fleming 1905, p. 480).

In any case, Fleming found an obvious and immediate use for the thermionic valve: rebuilding his connection to Guglielmo Marconi and the Marconi Company. Having filed a patent application for the valve, but not yet having experimented with it, Fleming wrote an undated letter to Marconi in which he mentioned his earlier invention of a wavelength-measuring instrument (the cymometer) and then remarked that his new device could "measure exactly the effect of the transmitter." He added: "I have not mentioned this to anyone yet as it may become very useful."[45]

Fleming's two inventions (i.e., the wavelength-measuring cymometer and the thermionic valve) evidently impressed Marconi, who expressed a wish to perform some experiments with the "electrical oscillation current rectifier." Fleming proposed setting up an apparatus at Marconi's factory at Chelmsford, where Marconi "might come to see it at some convenient time." In a Friday Lecture at the Royal Institution in March 1905, Marconi showed his approval by exhibiting Fleming's cymometer and valve. After the lecture, Marconi asked Fleming if he might borrow the valve for more experiments.[46]

Fleming took care not to miss this opportunity to re-establish his connection with Marconi. He replied immediately:

The only valve I have in my possession which works well is the one I lent you for your lecture, and with which all the work for my Royal Society paper was done. I do not wish to part with this valve as I shall have then no means of making comparison measurements. I will endeavour to get a couple of good valves made as soon as possible and sent over to you. Meanwhile I should like to draw your attention to the matter of my agreement with the Company about which I spoke to you when we last meet. In your absence I trust the matter will not be prolonged as it will be to the interest of the Company that it should be reestablished.[47]

In May 1905, the Marconi Company re-appointed Fleming as scientific advisor. The memorandum of agreement signed by the two directors of the company and Fleming contained four terms, the third of which specified that Fleming's inventions made between December 1, 1903 and May 1, 1905 would fall under the terms of the agreement as if he had been a scientific advisor during that time. Though the patent on the valve was issued in Fleming's name, the Marconi Company now had rights to it.[48]

Until about 1907, the valves were fabricated under Fleming's instruction in the workshop of the Ediswan Company, in Ponders End, by Charles Gimingham, a skilled lamp artisan and Fleming's lifelong friend. Whenever Marconi and the Marconi Company asked Fleming for valves, Fleming would order them and his assistant G. B. Dyke would fetch them. But *use* of the valves was not under Fleming's control even during this early period.

In late 1906, an Italian professor asked Fleming whether he could obtain a valve for scientific experiments. Thinking the request reasonable, Fleming told Marconi about it, adding: "Even if you wish to keep the exclusive use of the valve for wireless telegraphy, it might still be possible to supply it for such other scientific uses as do not conflict with your work."[49] Marconi, who was experimenting with the valve as a long-distance detector, refused. However, Marconi did decide to give several valves to the Italian government in 1907, and Fleming was asked to instruct the Italians in their use.

Marconi, not Fleming, transformed the thermionic valve into a sensitive detector for wireless telegraphy. In mid 1905, while experimenting with the device, Marconi found that it became much more sensitive when the galvanometer was replaced with a telephone. In 1906–07, he connected a telephone inductively to the valve circuit, using his oscillation transformer (the "jigger") as a means for this inductive coupling. With this modification the valve became "one of the best long distance receivers yet made, under some conditions better even than the magnetic detector." But it was still far from being a manageable device, for the signals became discernible when used with Marconi's new "continuous waves" (which were actually quasicontinuous). For example, Marconi, in his station at Poole, could receive clear signals sent from Poldhu with other receivers, but not with the valve, when a continuous wave was used.[50]

In 1908, Fleming devised a tungsten valve and a new circuit for connecting it to the receiver. Much as he had done in November 1904, Fleming wrote to tell Marconi "I am anxious this [improvement] should not be yet known to anybody but yourself." Marconi, however, did not welcome the new developments. After a few experiments, he informed Fleming: "Tungsten valves are not more sensitive than the best carbon valves . . . [and] your new arrangement of circuit does not give as good results as the standard circuit which . . . is strictly in accordance with the description given in my patent for the mode of using your valve as a receiver." After this, Fleming made no more important contributions to the valve's development, and Marconi and his young engineers began to take the initiative.[51]

During this early period, Fleming was apparently not concerned about the use of his valve by others. In October 1906, he told Marconi:

[The patent on the valve] is not by any means a strong patent and if we refuse to supply the valve for all purposes, people may import the similar device of Wehnelt from Germany or perhaps make it for themselves. I can hardly believe that the patent is worth fighting. . . . Personally I have no interest in the matter; I have already any little scientific credit there may be for the invention.[52]

A little later in October, the development of the thermionic valve took an unforeseen turn when Lee de Forest announced his invention of the audion at a meeting of the American Institute of Electrical Engineers. Basically, de Forest inserted a grid (applied to a separate potential difference) in the vacuous space between the plate and the filament. The current flow in the space was controlled by means of this grid. With this improvement, de Forest argued, the audion not only rectified but also amplified received signals.

De Forest tried to undermine Fleming's earlier contribution as much as he could. He asserted that Fleming's work on unilateral conductivity had been anticipated by the German scientists Julius Elster and Hans Geitel. He argued that the use of Fleming's valve was confined to "quantitative measurements over short distances." Concluding that "the value of [Fleming's valve] as a wireless telegraph receiver is nil," he argued that the audion was "tremendously more sensitive and available in practical wireless" than the valve (de Forest 1906a, pp. 748, 775).[53]

The abstract of de Forest's paper in *The Electrician* was enough to infuriate Fleming, who immediately countered that "the actual construction of the apparatus [the audion] is the same [as mine]" and that his valve had been actually used as a sensitive receiver in wireless telegraphy (Fleming 1906d). In reply, de Forest (1906b) again argued that the real genesis of his audion, as well as of Fleming's valve, was Elster and Geitel's research in the 1880s. De Forest (1907) then compared the valve to "a laboratory curiosity" and the audion to "an astonishingly efficient wireless receiver employing the same medium, but operating on a principle different in kind." Fleming's credibility was threatened, which was unbearable to him. This led him into an alliance of mutual interest with Marconi, since de Forest was one of Marconi's chief competitors. Fleming urged Marconi:

It is extremely important that de Forest in America should not be allowed to appropriate all methods of using the glow lamp detector. . . . I want our Company to have all the commercial advantage possible but I am anxious that de Forest shall not deprive me of the scientific credit of the valve invention as he is anxious to do.[54]

This was the beginning of Fleming's lifelong animosity toward Lee de Forest.

Fleming and de Forest again crossed swords in 1913 in the columns of *The Electrician*. But the controversy was not to be resolved in engineering journals; it was resolved in court. Since de Forest did not file a complete specification of his audion patent in Britain, his British patent on the audion lapsed. The Marconi Company, which held the patent rights to Fleming's valve, manufactured audions in Britain until 1918, when Fleming's patent expired. The Marconi Company tried but failed to extend the patent. In the United States, the Marconi Company sued de Forest for infringement of Fleming's valve patent. In 1916, a court in the United States ruled that the Marconi Company had infringed de Forest's patent and also that de Forest's audion was an infringement of Fleming's valve patent. A temporary arrangement was made between these two companies so that de Forest would produce the vacuum tubes and the Marconi Company would distribute them. But World War I, during which patent infringement was tolerated, made the arrangement unnecessary. It is interesting that in 1943, two years before Fleming's death, the US Supreme Court decided that Fleming's patent was "rendered invalid by an improper disclaimer" and had always been invalid, insofar as its first claim (in which Fleming did not distinguish between high- and low-frequency oscillations) contradicted the remaining claims (which concerned the valve's use in wireless telegraphy) (Howe 1944, 1955).

De Forest's audion apparently forced Fleming to ponder the originality of his own invention. He knew that the device itself—a lamp with one more electrode—had been made by Edison, as well as by Elster and Geitel. Rectification of high-frequency alternating current was also insufficient; it could be regarded as a "laboratory curiosity" with no practical importance. De Forest persistently argued that only his audion was suited to practical detection. Fleming, for his part, repeatedly emphasized that he had invented a new *detector* for wireless telegraphy, whereas all de Forest had done was add one more plate to his valve. This way of thinking seems to have been fixed in Fleming's mind after the patent litigation against de Forest.

Fleming's valve became more famous after the appearance of the oscillating audion in the mid 1910s. The audion began to be used not only as an amplifier but also as an oscillator for continuous waves. As the audion became more essential to radio engineering, the valve—as the audion's predecessor—was increasingly highlighted. However, Fleming was not always happy with the changed situation. In a sense, he became increasingly frustrated.

In 1918 the Marconi Company applied for an extension of Fleming's patent, but it was dismissed on the ground that the company had earned sufficient profits from it. However, Fleming later complained that he, "as the original inventor of it," had "never received a single penny of reward for it." In addition, in Fleming's view, the company had hurt his credibility by dissociating his name from his invention. Only after he resigned the scientific advisory to the Marconi Company did Fleming (1934, p. 147) express the opinion that "one firm has sold valves for many years made exactly in accordance with my patent specification, but which they advertise and mark 'Marconi Valves.'" This "injustice of some present-day commercial practice," he lamented, was the price he had paid for re-establishing his connection with the Marconi Company in 1905.

From Laboratory Effect to Technological Artifact

The transformation of a laboratory effect into a technological artifact, as a principal interaction between science and technology, has been of increasing interest to historians of science and technology (Hong 1994b; Wise 1988; Galison 1985; Van Helden and Hankins 1994).[55] The invention of the thermionic valve out of the Edison effect is a good example of such a transformation, and as such it has been interpreted in two different ways. The first interpretation highlights Fleming's scientific research on the Edison effect, such as his investigation of its mechanism in terms of electron theory, thereby providing a good example of the long-term contribution of scientific research to practical technology.[56] The second interpretation inverts this emphasis by highlighting the contribution of the Edison effect to scientific research. Simply put, it was the Edison effect, a curiosity that first emerged in a practitioner's workshop, that led Fleming (and others, including O. W. Richardson) to various scientific investigations of thermionic emission. The contribution of science to technology is less important here than the contribution of technology to science.[57]

The implications of these two interpretations are quite different. Those who wanted to praise Fleming's contribution emphasized his scientific research; those who wanted to disparage his originality argued that the valve was just an extension of the Edison effect. Similarly, those who wanted to underline the importance of scientific research for technology adopted the first interpretation; those who wanted to argue for the autonomy of technological development adopted the second interpretation.

Fleming is partially responsible for this conflicting situation, because he characterized his invention as resulting both from the momentous "happy thought" in 1904 and from his enduring scientific research in the 1880s and the 1890s. In Fleming's way of thinking, he potentially knew from his research that the Edison effect could be used for rectification; however, he did not carry out any investigations in the 1880s and the 1890s because of the effect's apparent uselessness at that time. His actual use of the effect came about in 1904, when technological and economic demand existed. Yet this account also relegates Fleming's role to a minor status, for here it appears as though the Edison effect, which showed a one-way flow of current, was destined to evolve into a rectifying device. Had Fleming not constructed the valve, someone else would have, for it was a simple matter to recognize in the Edison effect a means for building a rectifying device. One might say that a brutal "techno-logic" was inherent in the Edison effect, and that this irresistible logic, not Fleming's labor, produced the valve. (See figure 5.11.)

Yet, as we have seen, the thermionic valve emerged less from Fleming's enduring commitment to the Edison effect than from a manifold of specific

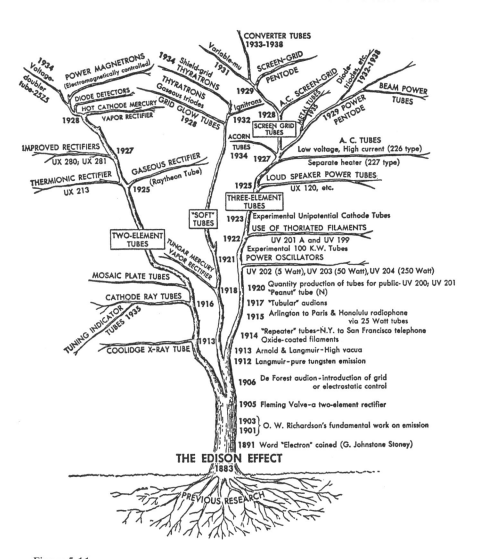

Figure 5.11
"The family tree of the thermionic tubes." Source: Maclaurin 1949, p. 91.

resources that were in place only in 1904. Some of these resources (such as the imperative of invention that he felt in 1904 to regain his connection to Marconi) pressured Fleming against his will; some (electron theory, the articles in *The Electrician*) were available to him as well as to others; some (photometric practice, the old lamp with an extra electrode) were unique to his laboratory; some (his interest in measurement and measuring instruments) came from Fleming's engineering research style. Some factors (such as Nodon's valve and Kelvin's paper) became significant only in the fall of 1904; others (the old lamp) had existed for a long time. Some factors were apparent to all concerned; others (such as Maxwellian difficulty with unilaterality) were hidden or even "unarticulated."[58] It was Fleming's dexterity in totality that manipulated these complex factors and eventually stabilized them into a useful artifact: the thermionic valve.

6

The Audion and the Continuous Wave

"What right have you to have that [audion oscillator] here? That thing is not yours. That belongs to Armstrong!"
—Michael Pupin to Lee de Forest, 1914 (de Forest 1950, p. 319)

In the 1910s it was discovered that the audion could be used as a feedback amplifier and as an oscillator (Tyne 1977; Aitken 1985, pp. 162–249). As an amplifier, the audion made it easy to receive feeble wireless signals; as an oscillator, it made the production of continuous waves, as well as the transmission of the human voice, simple and cheap. Wireless telegraphy was thus transmuted into radio.

The history of the audion is a human history—of engineers, scientists, businessmen, patent lawyers, amateurs, and their successes, their accidental discoveries, their misunderstandings, and their frustrations.[1]

At the center of this complicated human history was an artifact: the audion. Invented by Lee de Forest in 1906–07 to amplify wireless signals, the audion was a vacuum tube with a control grid inserted between the filament and the anode plate of J. A. Fleming's thermionic valve. De Forest apparently thought that the grid would control the flow of charged gaseous particles (which he called "ions") and amplify received signals as a relay did in telegraphy or telephony. De Forest was wrong: it barely amplified at all. Like Fleming's valve, the audion worked best as a rectifying receiver. However, between 1912 and 1914, several engineers found that it could be used as a feedback amplifier and (more important) as an oscillator producing continuous high-frequency Hertzian waves. These engineers soon filed a patent. Heated patent litigation began immediately. The first technical papers on the new features of the audion were published in electrical journals in 1914. By 1920 the importance of the audion was public knowledge.[2] The audion changed the theory and the practice of wireless telegraphy deeply and radically. In the 1920s it opened a new era of radio broadcasting.

The audion revolution[3] of the 1910s was an example of simultaneous innovation. At least four engineers, and perhaps six, arrived at the amplifying and oscillating audion circuits almost simultaneously. Edwin Howard Armstrong (1890–1954), an undergraduate at Columbia University, produced the circuits in September 1912; Lee de Forest (1873–1961), a flamboyant inventor, claimed to have invented virtually the same circuit in August 1912; Alexander Meissner of Germany made similar claims in March 1913, and Irving Langmuir (1881–1957) in 1913. The validity of their US patents was contested in what was known as the "four-party" interference proceedings. Fritz Lowenstein (in the United States) may have anticipated all the others early in 1912, and H. J. Round (in Britain) also devised and patented the amplifying circuit in 1913. The activities of these individual genius engineers were mingled with the strategies of large corporations. Meissner worked for the German firm Telefunken; Langmuir was a well-known scientist-engineer at General Electric; Round was the foremost engineer at the British Marconi Company. AT&T bought de Forest's audion patents; Westinghouse later bought Armstrong's; Lowenstein provided much information to Western Electric and GE.

Why did these six engineers discover the new features of the audion only after 1912, though the device had been invented in 1906–07? How is it that they almost simultaneously devised the audion amplification and oscillation circuits? Were they working independently, or did each know of the other's work? One might argue, as some historians have, that this is an instance of technology's having "its own life" (Aitken 1985)—that the audion inherently embodied amplification and oscillation.[4] However, these features were not revealed until after 1912. There is something to be said for this way of thinking, because no one, including the audion's inventor, expected in 1906 that the tiny lamp might possess amplifying and oscillating features. In this sense, no one effectively controlled or predicted the audion's future.

This chapter has two aims. The first is to provide an explanation of how de Forest invented the grid audion at the end of 1906. In particular, I will critically discuss how Fleming's valve influenced de Forest's audion. The second aim is to present the audion revolution as a path of continuous development from the "negative resistance" of the electric arc to the arc generator to the audion. I will reveal a theoretical and instrumental continuity between the arc and the oscillating audion, a continuity that was most apparent in that each hissed or whistled. This continuity provides a proper context for the central question posed above: How did several engineers, in 1912–1914, almost simultaneously transform the rectifying audion into a device for amplification and oscillation?

FitzGerald's Solution to Continuous Waves: Negative Resistance

We first go back to Heinrich Hertz. As we saw in figure 1.1, Hertz's spark transmitter, the device that first produced artifactual electromagnetic waves, consisted in essence of two condenser plates, a straight wire connecting them, a spark gap in the middle of the wire, and an induction coil that supplied energy to the condenser plates. The induction coil charged the condenser plates to a high voltage until the air in the wire gap became temporarily conducting. At this point, the plates, the wire, and the gap constituted a single linear oscillator. As the oscillator lost its energy in the form of electromagnetic waves, the air gap eventually ceased conducting, and the process was repeated. We saw in chapters 1 and 4 that the resultant wave was highly damped, precisely because the Hertzian device did not continuously secure energy from the source (Buchwald 1994). Owing to damping, some Maxwellians reasoned, the distance traversed by the wave was too short for any practical purpose. Hertz and others were more concerned that, because of damping, the wave had a broad spectrum, and therefore many different resonators would respond to it—making tuning impossible (figure 6.1). According to the consensus that had emerged sometime in the mid 1890s, the damped wave was effectively a mixture of many waves with different frequencies.

Hertz's transmitter may be regarded (approximately) as a circuit consisting of capacitance (C), inductance (L),[5] and resistance (R). In the case of Hertz's transmitter, the resistance R of the oscillatory circuit consists of both the ohmic resistance of the wire and the "radiation resistance" $(R = R_{ohmic} + R_{radiation})$, the latter being much greater than the former. We can easily see the effect of resistance (whatever it is) on the circuit by considering a simple L-R-C circuit,[6] the governing equation for which was well established

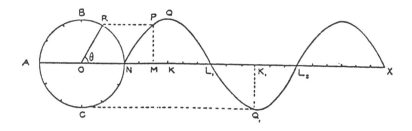

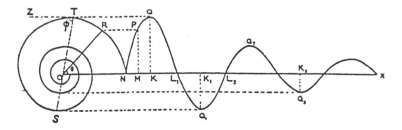

Figure 6.1
Continuous (above) and damped waves.

in the 1890s. The total voltage drops across L, R, and C must be zero for an isolated circuit:

$$\frac{d}{dt}(LI) + Ri + \frac{1}{C}\int i = 0.$$

Assuming that L is time independent, the above equation is again differentiated with respect to t:

$$\frac{d^2i}{dt^2} + \frac{R}{L}\frac{di}{dt} + \frac{1}{LC} = 0.$$

Solving this, we find

$$i - Ae^{-(R/2L)t}\sin\omega t,$$

where

$\omega = 1/\sqrt{LC}$.

The graph of the current i is precisely like the lower graph in figure 6.1. This solution also shows that damping is determined by the factor $e^{-R/2L}$ (see also chapter 4). The larger the resistance R, the heavier the damping. If R (= $R_{ohmic} + R_{radiation}$) is zero, there is no damping.[7]

One might minimize ohmic damping caused by the ohmic resistance by using a substance with the least possible resistance, but one could not eliminate radiation damping caused by the radiation resistance; in fact, it was precisely because of the radiation resistance that radiation occurred at all. It was not possible to substantially diminish resistance, which derived mostly from the transformation of circuit energy into the energy of electromagnetic waves. Producing continuous waves therefore seemed practically impossible, at least with a linear oscillator. By changing the shape of Hertz's device, physicists and engineers could produce less damped waves. For example, waves produced by a circular oscillatory circuit (suggested by Lodge) were less damped than those produced by Hertz's linear antenna. But the circular oscillator produced weaker waves, since the energies involved were about the same as with Hertz's device. Struggling to solve this problem, Marconi ingeniously combined a closed oscillatory driving circuit with an open radiating antenna. His syntonic system, represented by the "four-seven" patent, produced less damped, more powerful waves. Yet one should not forget that the wave produced by Marconi's "four-seven" syntonic transmitter was still damped, since the driving oscillator was periodically recharged by the primary of an interrupter-driven induction coil. As tuning emerged as the central technical issue for wireless communication, the Marconi Company used every method at hand to produce less and less damped waves, including the use of Marconi's specially constructed disk dischargers.[8]

To obtain true continuous waves, one had to eliminate the effect of the resistance of the oscillatory circuit. The physicist George FitzGerald was the first to offer this solution. As early as late 1891 FitzGerald had wondered whether he could use Hertzian waves for lighthouse-to-shore communication (see chapter 1). He was surprised to find that the maximum transmission distance of Hertzian waves was too short to be useful for any practical purpose. Damping, FitzGerald reasoned, was a major reason for this. He compared the wave emitted from Hertz's spark transmitter to the sound of a cork popping out of a bottle. It only existed for a very short time—a hundredth or even a thousandth of the time interval between two adjacent sparks. In effect, the energy was more like a pulse than a continuous emission. It was not sufficient to continuously stimulate the resonator at a distance. What was needed was something like an "electric whistle"— something that could produce a strong beam of continuous waves, just as an ordinary whistle produced continuous sound (FitzGerald 1892).

FitzGerald suggested two different methods for producing continuous waves. The first would require the construction of a huge AC dynamo that would generate "a frequency of about one million [Hz]." However, FitzGerald knew that this was a dream; such a high speed would break an ordinary dynamo into pieces. FitzGerald's second method (which was at least possible) consisted of connecting an ordinary dynamo in series with an L-R-C circuit and obtaining the discharge from this condenser-dynamo hybrid circuit. FitzGerald's explanation of this circuit was brief. He pointed out that since the dynamo rotates, the inductance of the resulting circuit was not time-independent. Let us go back to the circuit equation

$$\frac{d}{dt}(LI) + Ri + \frac{1}{C}\int i = 0.$$

Previously, we assumed that inductance L does not vary with time. But if it does, the equation becomes

$$L\frac{di}{dt} + \left(R + \frac{dL}{dt}\right)i + \frac{1}{C}\int i = 0.$$

FitzGerald (1892) succinctly commented:

Calling the quantity of electricity on the condenser Q [$i = dQ/dt$], the differential equation for a dynamo of inductance L and resistance r and a condenser of capacity X, is

$$L\ddot{Q} + (r + \dot{L})\dot{Q} + \frac{Q}{X} = 0.$$

If dL/dt be $= 0$, the solution of the equation is $Q = Q_0 \, e^{(-rt/2L)} \cos\omega t$, and the rate of degradation of amplitude depends on the factor $e^{(-rt/2L)}$. If, however, $-dL/dt$ be greater than r the exponent of e becomes $+$ and hence Q would go on increasing.

In this case, the self-induction of the dynamo has the effect of canceling out the resistance of the circuit. A circuit with effectively zero or negative resistance could thus be made. This would have the consequence of constantly accumulating charge Q to overcome the effect of the resistance (ohmic plus radiation) that caused damping. As Lodge (who chaired the Physical Society's session at which FitzGerald read his paper) sensed, the fact that "the damping factor could be changed in sign must have tremendous consequences."[9]

But theoretical principles and working technology are often quite different. FitzGerald (1892) reported that he "tried experiments with Leyden jars (i.e., condensers) and a dynamo, but got no result." What was needed was not a real dynamo but, as FitzGerald later said, a sort of "*molecular dynamo*" that could cancel out the resistance of the main oscillatory circuit. But where could one find such a molecular machine?

The Ayrtons, Duddell, and the Arc Generator

In the early 1890s, over 3 years, William Ayrton, a professor of physics and electrical engineering at the Central Institution in London, performed with his students a series of painstaking experiments on the voltage-current relationship of the electric arc. These experiments were difficult and time consuming because of the tricky and unstable nature of the arc. In 1893, Ayrton read a paper on the experiments at the Electrical Congress in Chicago and submitted it for publication in the proceedings of the congress. The paper was accidentally burned; not even the abstract survived.[10] Extremely busy with other professional activities as an expert engineer, Ayrton did not want to repeat the time-consuming experiments. At this juncture, his wife Hertha volunteered to perform the experiments. She asked her husband if she could use his laboratory and obtain help from his college students at the Central Institution, and William Ayrton agreed to provide what she wanted.[11]

The results of Hertha Ayrton's experiments were published from early 1895 on as a series in *The Electrician*. The papers were quite detailed, including many graphs and fine drawings. One of the graphs, reproduced here as figure 6.2, shows that, as the current or voltage varies, an electric arc goes through two different stages; silent during the first, it hisses during the second. The current of the hissing arc was particularly anomalous and unstable, and its sound often changed from hissing to whistling and even howling. Strangely, as figure 6.2 shows, the voltage remained inversely proportional to the current during both stages. That the electric arc violated Ohm's Law had been known to scientists since the late 1860s, but the

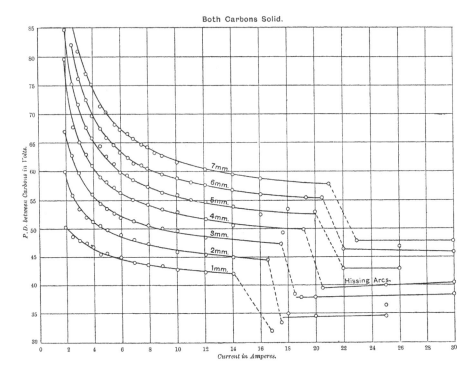

Figure 6.2
Hertha Ayrton's measurement of the current-voltage characteristic of the electric arc. Original legend: "P.D. and Current for Different Lengths of Arc. Carbons:— Positive 11mm., negative 9mm." Note the "negative resistance" in the "hissing unstable state." Source: Ayrton 1895. p. 337.

exact relationship between current and voltage had been controversial. Hertha Ayrton proposed her own formula. In so doing, she challenged Silvanus Thompson, an authority on many topics in electrical science and engineering who had given a Cantor Lecture on the electric arc before the Royal Society of Arts just a few months earlier. Hertha Ayrton argued that the formula that Thompson had given was incorrect, and that it should be replaced by a more sophisticated formula of her own devising (Ayrton 1895, pp. 638–639).[12] Thompson ignored the challenge. However, he could no longer ignore it when William Ayrton, on the basis of his wife's experiments, argued that the "true resistance" of an electric arc was negative.

To William Ayrton, the resistance of an arc was defined in terms of the ratio of the change in the arc's voltage to the corresponding change in its current. Hertha Ayrton's graph showed that the voltage decreased as the current increased. If the ratio dV/dA was the true resistance, it had to be negative. To Silvanus Thompson, however, "negative resistance" had no meaning or was at best a joke. To Thompson, resistance was analogous to friction and thus negative resistance was analogous to negative friction. Since resistance, like friction, created heat, energy conservation required that negative resistance cause heat to be absorbed from the environment. This was simply absurd, Thompson reasoned, because it violated the principle of the second law of thermodynamics. Instead, Thompson asserted, if the resistance was to vary inversely with the current, an "effect" akin to what might be called a negative resistance would appear (Ayrton 1896a,b; Thompson 1896). *The Electrician* agreed that negative resistance was as heterodox a notion as negative friction: "Resistance is the youngest member of the electrical trinity, and we have learned to regard it as the most tangible."[13]

Few physicists were interested in this debate. Not many scientists were familiar with the strange property of the electric arc. Asked for his opinion by *The Electrician*, Oliver Heaviside (another authority on electrical science and engineering) said simply "I do not know enough about the arc; I prefer gas lighting for personal use."[14]

George FitzGerald jumped into the debate and criticized Thompson. FitzGerald opined that negative resistance was feasible because it did not violate any fundamental principles of physics. Negative resistance never existed alone; a larger positive external resistance was always needed to make the arc glow. Therefore, the electric arc light as a physical system would not violate the second law of thermodynamics in spite of the arc's localized negative resistance. FitzGerald (1896b) also suggested a new way to understand negative resistance: The usual voltage drop across an ordinary (positive) resistance could be regarded as a kind of "back emf" (or counter emf) that caused the drop. If, contrary to the back emf, another emf caused a voltage gain, that emf could be counted as a negative resistance. This was similar to having a running dynamo in the circuit, helping to drive a current through it. In fact, such an emf had previously been discovered in the Thomson effect (a thermoelectric phenomenon in which the generation or absorption of heat occurs where an electric current flows across a metal in or against the direction of the metal's temperature gradient) and had been named the "adjuvant emf." "These two cases (a dynamo and the Thomson effect)," FitzGerald wrote, "have many points in common [because] there is very little essential difference between a dynamo turning electromagnetic energy into heat by friction on a large scale and the innumerable molecular machines." The negative resistance of the arc had the effect of behaving like a dynamo continuously pumping energy into the circuit.

If, as FitzGerald claimed, negative resistance really existed in the arc, then the arc could be utilized for the production of continuous waves. This was in fact accomplished by William Ayrton's favorite student, William Duddell. According to one obituary, at the age of 4 years Duddell had transformed a toy mouse into an automaton by combining it with a clockwork. This story may be hard to believe, but Duddell was truly an exceptionally ingenious instrument maker. For example, he was the inventor of the first practical oscillograph, an instrument used to photograph AC waveforms. The problem of building such an instrument had escaped the powers of many eminent scientists and engineers. Using this device, Duddell challenged the transformation of the arc into a high-frequency oscillator. He combined the hissing arc's negative resistance with an ordinary *L-R-C* oscillatory circuit to pump energy, thereby canceling out the resistance of the *L-R-C* circuit. The trick was to use the oscillograph to find the specific conditions that would stabilize this circuit. In 1900, Duddell demonstrated a reliable device for that purpose: the singing, or musical, arc. Duddell's arc (figure 6.3) indeed succeeded in producing continuous waves, but only in the audio range (about 15,000 Hz)—hence the name. Duddell used it to play "God Save the Queen" at a meeting of the Institution of Electrical Engineers. This drew much attention, but neither Duddell nor any anyone else attempted to increase the frequency to the region of electromagnetic waves (around 10^6 Hz).

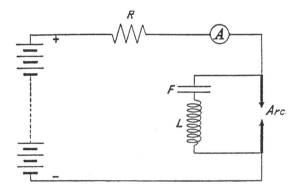

Figure 6.3
Duddell's singing arc. Source: Duddell 1900.

Duddell's singing arc did, however, inspire other engineers. A Danish physicist, Valdemar Poulsen, constructed an arc generator that produced continuous waves in the megahertz range. The purpose of the Poulsen arc (which was the culmination of several years of research by Poulsen and P. O. Pedersen) was to increase the maximum frequency of Duddell's singing arc. Poulsen and Pedersen first found that the arc produced oscillations of much higher frequency in hydrogen gas than in air. They then discovered that a magnetic field across the arc would increase its frequency, and that making the positive electrode of water-cooled copper instead of carbon made the arc more stable in increasing its frequency. Finally, they rotated the carbon electrode, which was crucial for the arc's stability. These four elements—the hydrogen gas, magnetic field, water-cooled positive copper electrode, and rotation of a carbon electrode—constituted the Poulsen arc.[15] In 1907 the Poulsen arc generator began to be used to produce continuous waves for wireless telegraphy and telephony.[16]

The works and devices of Duddell and Poulsen made it widely known that devices having negative resistance could be employed to produce continuous waves. Since the arc generator was patented by Poulsen, other engineers built negative resistance into different devices. Two Germans, Ernst Ruhmer and Adolf Pieper, substituted a mercury lamp for the arc. The lamp had been known to have the characteristic of negative resistance. They connected the lamp in parallel with an oscillatory circuit for "generating permanently undamped oscillations." Frederick Vreeland, an American, also applied for a patent for the use of a mercury vapor converter for the production of continuous waves. A Russian, Simon Eisenstein, used a cathode ray tube for the same purpose.[17] (See figure 6.4.) Negative-resistance devices were also used in telephony. Telephone engineers found that the mercury lamp worked well as a repeater (a relay) to amplify telephonic signals. The physical nature of "negative resistance" remained mysterious, but the lack of theoretical understanding did not prevent engineers from using the effect in their daily practice.

Marconi and Fleming on the Poulsen Arc

Around 1905, the production of continuous waves became the central issue in wireless communication, for two reasons. First, in wireless telegraphy continuous waves could secure precise tuning and therefore secrecy. Second,

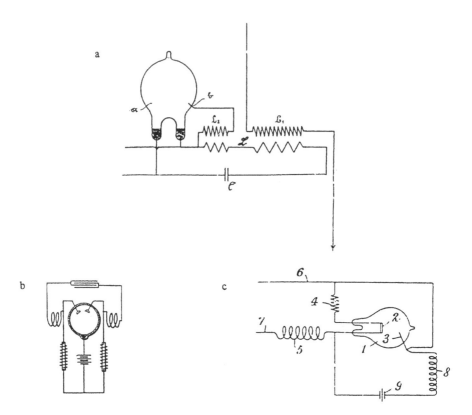

Figure 6.4
Various circuits utilizing negative resistance to produce undamped oscillations. (a)
Ernst Ruhmer and Adolf Pieper German patent 173,396 (1904). (b) F. Vreeland,
US patent 829,934 (1906). (c) S. Eisenstein, US patent 921,526 (1909).

and more important, they made wireless telephony possible. Damped waves
could not carry a human voice, but continuous waves could. Marconi had
nearly monopolized important patents on spark telegraphy—patents on
antenna connection, inductive tuning (the "four-seven" patent), the coherer,
the magnetic detector, and Fleming's valve—but had hardly any for continuous waves and wireless telephony. This was Marconi's Achilles' heel.

Nevil Maskelyne, one of Marconi's enemies, introduced the Poulsen arc to
Britain in 1906. Around the same time, Maskelyne helped Lee de Forest to
build the British De Forest Company. With the rights to exclusive use of the
Poulsen and de Forest patents in Britain, Maskelyne gathered a number of
opponents of Marconi and formed the Amalgamated Radio-Telegraph

Company, with Lord Armstrong as chairman and Maskelyne himself as scientific advisor. In November 1906, Maskelyne demonstrated the Poulsen arc publicly and announced that he had communicated across 530 miles with a 1-kilowatt arc generator. He also predicted that transatlantic communication should be readily obtained with a power of only 10 kilowatts. This, William Preece commented, "sounded the death knell of spark telegraphy."[18]

The Poulsen system was the first practical system that was radically different from the Marconi system, in that the spark was completely dispensed with. Some engineers believed that the Poulsen arc would ultimately lead to wireless telephony, which at the time had only reached the stage of laboratory experiment. The immediate value of the Poulsen arc lay in precise tuning. As *The Electrician* commented, "the chief advantage of the new method, or of any method producing a continued electrical monotone, is that syntonic working will now be able to live up to its name."[19] Though it was not immediately usable for wireless telephony, its precise syntony and ability to transmit 530 miles generated a great deal of interest among wireless engineers.

Against Maskelyne's claims, the lead article in *The Electrician* of December 21, 1906 cast serious doubt on the efficacy of the Poulsen arc. I suspect that this article was written by, or with the help of, J. A. Fleming.[20] After pointing out the existence of unilateral conductivity in the Poulsen arc, its author emphasized that the Cooper-Hewitt mercury rectifier with unilateral conductivity proved to yield a "very rapidly intermittent series of damped oscillations" instead of truly continuous ones. The article also pointed out that the use of the magnetic field in the Poulsen arc "reminds us of the older blow-out methods" used for rapidly intermittent current. From these considerations, the author drew a conclusion that the Poulsen arc produced "regularly and rapidly intermittent short trains of waves" instead of "continuous trains of undamped electrical oscillations." In addition, he maintained, the arc was more cumbersome and troublesome than the spark, because it required a separate DC generator. For these reasons, "it may yet be many years before the musical arc can sing the dirge of the spark."

Though skeptical, Fleming did not ignore the new system of wireless telegraphy. On the first day of 1907 he wrote a long letter to Marconi addressing various issues, including the Poulsen arc. An excerpt follows:

There is no doubt that many difficulties will have to be overcome before the Poulsen arc becomes as simple and easy to handle as the spark. For reasons too long to enter into, there is a great difficulty in maintaining an electric arc without constant interruption in hydrogen and in a strong magnetic field, and of course unless this can be done so that the arc will work quietly and regularly for hours together its practical and commercial use is very seriously discounted. No doubt a number of very pretty experiments can be shown with continuous oscillations and a remarkable sharp tuning, but from the practical point of view we have to consider a number of quite different problems. On the other hand in a letter sent to Mr. Hall before Christmas I urged that Poulsen's invention was not a thing to be entirely depreciated because of the invention to which it may lead. . . . still I think it would be only prudent to examine Poulsen's method most carefully both in the Laboratory and in practice, and I am very glad to hear that you have methods for producing continuous waves and also that you agree that we should look into the question of high frequency alternator.[21]

"Methods for producing continuous waves" meant Marconi's disk discharger method, which was then evolving. In 1903, Fleming had devised a way to prevent erosion of the two balls in the discharger by having them rotate. When Marconi tried replacing the slowly rotating balls with three rapidly rotating disks, he found that highly persistent 0.2-megahertz oscillations were produced. Marconi seemed to think this effect to be "neither an oscillatory spark or an ordinary arc." But there were problems in detection. In modern terms, there was no audio-frequency modulation in receivers for telephone detectors, so the making and breaking of a Morse key in a transmitting station were not easily recognizable on the telephone in the receiving station. Marconi initially solved this by connecting and disconnecting the telephone from the aerial by means of a revolving interrupter, but he soon found that copper studs fixed around the periphery of the middle disk occasionally interrupted the continuous wave, which was audible on the telephone receiver.[22]

Fleming experienced numerous troubles with the Poulsen arc. Throughout his experiments in the Pender Laboratory, he found it very difficult to keep the arc stable for use in the field. Any tiny variation in conditions caused the arc to fluctuate significantly or to be extinguished. In addition, the carbon deposits inside the vessel from the heated coal gas caused problems. Fleming concluded that the Poulsen arc was still a laboratory instrument, not a practical telegraphic device. "If you can perfect your methods sufficient to give a high frequency, high enough for practical purposes," he assured Marconi, "I think it will be superior to Poulsen's method for a large power station."[23]

Fleming gave a series of public demonstrations to show that the Poulsen arc was far from a practical device. He claimed that it produced intermittent, not continuous, oscillations. By connecting a long resonance coil to the Poulsen arc and rapidly moving a neon tube along the coil, he showed that the band of light on the tube was cut by several dark lines. Similarly, when the tube was made to rotate near a helix-coil detector, it did not give a uniform disk of light; it gave several dark bands (Fleming 1907b, p. 692).[24]

Using his valve (and a telephone) as a detector, Fleming showed that "the telephone yields a sound which shows that the continuous current through it is interrupted irregularly, and this can only be because the oscillations in the arc circuit are interrupted" (1907c). And the Poulsen system had other problems. When compared with Marconi's feebly damped oscillations, Poulsen's putative undamped waves were not really superior to damped waves. There was no difference in transmitting distance and tapping.

In regard to tuning, Fleming reported that he was "still awaiting quantitative confirmation," since tuning was dependent on the kinds of detectors used. Moreover, the power required for ordinary ship-to-shore transmission across 200 miles was 0.2 horsepower for spark telegraphy, but 1–1.3 horsepower for the Poulsen arc generator. In view of these factors, it seemed that "for short distance work . . . the spark method has advantages denied to the arc."[25]

The Poulsen arc, introduced in England by the "scientific hooligan" Nevil Maskelyne, proved technically problematic. Fleming's experiment on the arc in 1907 confirmed his and Marconi's anticipation and fulfilled their expectations.

In the following few years, the Marconi Company tried to perfect its spark telegraphic system. Marconi's disk discharger produced almost continuous waves. It was, however, the pinnacle of an old technological regime, rather than the beginning of a new one. The new technological regime—a technological revolution—originated neither from spark nor arc. It began with a tiny lamp, the audion.

De Forest's Audion

The three-electrode (or grid) audion was invented by the American engineer Lee de Forest in late 1906. De Forest had earned a Ph.D. degree at Yale University in 1899 with research on the reflection of Hertzian waves at the end of wires. He then went into the business of wireless telegraphy. In 1901, as was mentioned in chapter 4, he competed with Marconi during an international yacht race; his signals interfered with Marconi's. In 1902 he became the vice-president and director of the American De Forest Wireless Telegraph Company, which built stations and sold stock but sent very few messages. De Forest's role in this company was to boost the sale of stock by creating a public sensation. The company closed operations in 1906. In that same year de Forest invented the audion, which he would later call the "Aladdin's lamp of our new world" (de Forest 1950, p. 1).

The De Forest Radio Telephone Company, which sold audions and other devices, was formed in 1906 and went bankrupt in 1911. In 1912 the US Department of Justice charged de Forest and his associates with having used fraudulent methods to promote the company, "whose only assents were de Forest's patents on a strange looking device like an incandescent lamp which he called an audion and which had proven worthless." De Forest narrowly avoided jail.[26]

De Forest was, he later recalled, eager to invent a sensitive new detector. Marconi had used and patented the coherer and the magnetic detector, and the American engineer Reginald Fessenden had patented the electrolytic detector. While working on an electrolytic detector in 1900, de Forest accidentally discovered that a gas burner in his room responded to sparks. Though he soon discovered that the gas burner's response was caused by the sound of sparks, not by electromagnetic oscillations, this was the starting point of his research on the use of gases as a detector of Hertzian waves.

De Forest's first detector, which eventually led to the three-electrode audion of 1906, utilized the flame of a Bunsen burner. Why a flame? It was known that the flame of the Bunsen burner did not follow Ohm's law (de Forest 1906a, pp. 737–738). In fact, like Fleming's valve, a flame exhibits unilateral conductivity. De Forest, however, attributed this strange effect to the action of ionized gases, and, in particular, to the difference in the velocity between positive and negative ions (i.e., electrons). He reported the following (de Forest 1906a, p. 738):

. . . on account of the ionization of the gas near the incandescent metal, and the greater velocity of the negative over the positive ions, it is to be expected that even if no external electromotive force be applied to the electrodes . . . a current will pass along a wire connecting the two electrodes, whose direction is negatively from the hotter to the cooler body in the flame. . . . Now if the Hertzian oscillations traverse the hot gas, the momentary potentials thereby impressed upon the moving ions will conceivably interfere with motions or with the rates of recombination between positive and negative ions, and thus affect the current through the wire.

When an external voltage was applied to two points within a flame, the current between these two points first increased with the increase in applied voltage, then ceased to increase, then began to increase rapidly again when the applied voltage reached at a certain point. De Forest found that the external voltage determined the sensitivity of the "flame Audion." The flame detector was as sensitive as electrolytic detectors; however, when "considerable difficulty was found in getting an absolutely steady flame," de Forest moved to try other substances, such as the electric arc (ibid., p. 740).

Although de Forest later recalled that he had performed experiments on flame detectors in 1903, the first patent on them (US patent 979,275) was written November 4, 1904 and filed February 2, 1905. This patent concerned six devices (figure 6.5) whose general operating principles were as follows:

The separation between them [the electrodes] may be neutralized sufficiently to enable them to act as a detector of electrical oscillations, if the intervening surrounding gaseous medium be put into a condition of molecular activity, such for instance as would be caused by heating it in any manner, as by radiation, conduction, or by the combustion of gases in the space which surrounds the pole; such condition or molecular activity causes what would otherwise be a non-sensitive device to become sensitive to the reception of electrical influences.

What de Forest meant by "molecular activity" was probably "ionization," as he remarked that ionization was "more or less accomplished or greatly facilitated by their previous heating which has already put them into a condition of intense molecular activity."

What was the effect of Hertzian oscillations, and how could they be detected? In describing the action of the flame detector represented by Fig. 1 in figure 6.5, de Forest stated that electrical oscillations seemed to "break down or lower the insulating quality of the gap," which made the current from battery *B* flow through a local circuit to affect telephone *T*. However, in describing Fig. 5 he wrote: "The oscillations apparently ionize the gas and thus temporarily reduce its insulating power."[27]

When de Forest filed a patent for his flame detectors, he could not have known of the thermionic valve, which Fleming had invented and for which he filed for a patent in November 1904. Therefore, the similarity of one of de Forest's flame detectors (Fig. 6 in figure 6.5) to Fleming's valve is interesting. Even its circuit diagram resembles the valve circuit. A closer look at the device, however, reveals differences. De Forest's Fig. 6 is a linear descendent

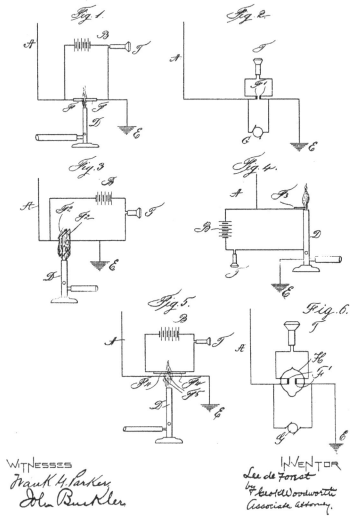

Figure 6.5
Lee de Forest's early (1904) flame detectors.

of Fig. 2. In both Fig. 2 and Fig. 6, *G* is a dynamo for heating the gas between two electrodes F^1. The only difference between Fig. 2 and Fig. 6 is that in the latter the electrodes are enclosed in the receptacle *H*. Both figure show flame detectors in which hot electrodes increased the molecular activity of the gas between them. De Forest believed that the electric oscillations increased the gas's conductivity and allowed current to flow in the local telephone circuit.[28] The receptacle *H* was helpful in that it made the gaseous medium stable, but it was not essential. De Forest later filed a separate patent on the "receptacle device" (Fig. 6), but he did not provide details on its construction.[29]

De Forest's next important patent (US patent 824,637, filed January 18, 1906) employed lamps with filaments (which, according to him, "may be ordinary incandescent-lamp carbon filaments").[30] The patent includes claims for six devices (figure 6.6). In both conceptual and material senses, these "filament devices" are a continuation of de Forest's work on the "receptacle device." His emphasis on the molecular activity of the gaseous medium remained unchanged. As before, the patent contained a passage describing "a receptacle inclosing a sensitive gaseous conducting medium" without giving specific details on it.

De Forest's filament devices were the predecessors of a new family of devices. De Forest soon transformed them into the two-electrode audion, and within a year he invented the three-electrode (grid) audion. What motivated him to shift his attention from the "receptacle device" to the "filament device"? He later recalled that his study of the incandescent filament was motivated by his realization that it had features in common with a gaseous flame (de Forest 1920, pp. 4–5):

It was not until 1905 that I had opportunity and facilities for putting to actual proof my conviction that the same detector action which had been found in the neighborhood of an incandescent platinum wire in a gas flame existed also in the more attenuated gas surrounding the filament of an incandescent lamp. In one case the burning gases heated the electrodes; in the other the electrodes heated the gases. But in both it was, *first,* the electrons from the hot electrodes, and, *second,* ionization of the gases which these electrons produced, that established an electrically conducting state which was extraordinarily sensitive to any sudden change in electrical potential produced on the electrodes from an outside source.

In the previous chapter, we saw how much de Forest's audion upset Fleming. The two men fought in the columns of *The Electrician* and later in court. The crucial question here is how much Fleming's valve influenced

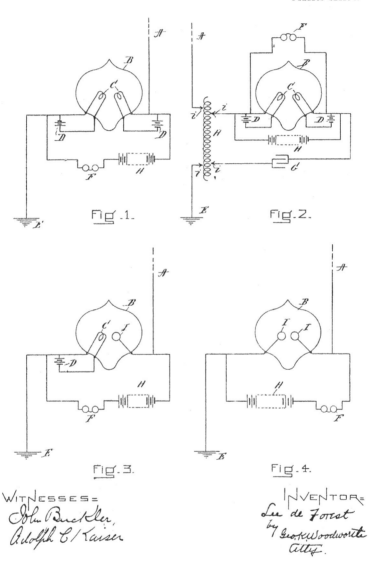

No. 824,637.   PATENTED JUNE 26, 1906.
L. DE FOREST.
OSCILLATION RESPONSIVE DEVICE.
APPLICATION FILED JAN. 18, 1906.

2 SHEETS—SHEET 1.

Fig. 1.   Fig. 2.

Fig. 3.   Fig. 4.

WITNESSES=
John Buckler,
Adolph E. Kaiser

INVENTOR=
Lee de Forest
by Geo. K. Woodworth
Atty.

No. 824,637.
PATENTED JUNE 26, 1906.
L. DE FOREST.
OSCILLATION RESPONSIVE DEVICE.
APPLICATION FILED JAN. 18, 1906.
2 SHEETS—SHEET 2.
B
A
I
J
F
M
L
E
Fig. 5.
A
F
J
B
O
P
Q
K
E
Fig. 6.
WITNESSES:
John Buckler,
Adolph C Kaiser
INVENTOR:
Lee de Forest
by Geo. K. Woodworth
Atty.

Figure 6.6
Lee de Forest's "filament devices."

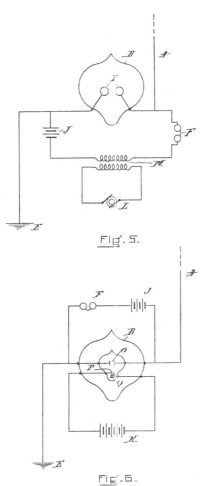

de Forest's invention (if at all). De Forest did not deny that he had known of Fleming's work. Indeed, he explicitly stated the following: "At the period now under consideration, 1903–05, I was familiar with the Edison effect and with many of the investigations thereof carried on by scientists, Prof. Fleming among others." (de Forest 1920, pp. 4–5) On October 26, 1906, speaking on the (two-electrode) audion at a meeting of the American Institute of Electrical Engineers, de Forest also mentioned Fleming's valve and its rectifying effect, though his purpose was to highlight the difference between them (de Forest 1906a, p. 748). However, he did not admit that Fleming's valve had had any direct, crucial influence on his invention of the audion. Consider the following assertion (ibid.):

In 1904 I had outlined a plan of using a gas heated by an incandescent carbon filament in a partially exhausted gas vessel as a wireless detector, in place of the open flame. But here the rectification effect between hot filament and a cold electrode was not considered. Two filaments, heated from separate batteries would give the desired detector effect equally well. What I had already found in the flame detector, and now sought in a more stable and practical form, was a constant passage of electric carrier in a medium of extraordinary sensitiveness or tenuity, which carriers could be in any conceivable manner affected to a marked degree by exceedingly weak electrical impulses, delivered to the medium, indirectly or through the hot electrodes.

This assertion is not entirely correct. As we have seen, de Forest's "plan of using a gas heated by an incandescent carbon filament" was not formed in 1904, but in January 1906[31]—when Fleming's valve was already well known.

Although de Forest never admitted to Fleming's direct influence in 1905–1907, evidence of such influence can be found in one of his patent specifications. In December 1905, just before filing the patent for the filament device, de Forest filed another for the "static valve for wireless telegraphy," a device for preventing static in the receiving antenna. Fleming's valve was clearly used in de Forest's device. De Forest noted in the patent that "the device V^1 . . . is an asymmetric resistance or electric valve which has been fully described by J. A. Fleming in a paper published in the *Proceedings of the Royal Society of London*, March 16, 1905."[32] This does not, however, lead to the conclusion that de Forest replicated Fleming's valve, or that de Forest's originality was nil. As we have seen, de Forest's filament device had part of its genealogy (in both theoretical and material senses) in the receptacle device. Yet it is also true that by the end of 1905 de Forest knew about the valve and had experimented with it, although in recollecting the audion's birth he minimized its influence. In addition, de Forest probably could have discovered that the valve, or the incandescent lamp, was more stable and practical than his receptacle device. This would have been a stronger motivation for his shift of attention from flames to the incandescent lamp in late 1905 or early 1906.

Of the six devices for which de Forest applied for patents in January 1906, the third (Fig. 3 in figure 6.6) was the one most similar to Fleming's valve.[33] However, there were differences. De Forest still believed that his device utilized ionization—a principle quite different from the valve's principle of rectification. A sudden increase in the gas's conductivity due to oscillations, he believed, temporarily completed the local battery circuit, allowing current to flow through the telephone. In this sense, de Forest believed, the device acted like a relay, or a trigger device: "When an independent external source of electromotive force is applied, it then operates as a *relay* to the Hertzian energy instead of merely rectifying this energy so that it can be used directly to give the sense signal." (de Forest 1906a, p. 748)[34] To him, this was good enough as a fundamental difference between his device and Fleming's valve. Even in 1920, he asserted: "The two electrode audion, with A and B batteries, was not primarily a 'valve.' And I have always objected to this misapplication of the name valve to the audion; a name which our British friends have from the first persisted, with a stubbornness worthy of a better cause, in misapplying!" (de Forest 1920, p. 6)[35]

After filing his patent in January 1906, de Forest traveled to Europe to advertise his company's transatlantic venture and to sell stock. A few months later, he had to take a break from some experimental work he was doing in Ireland and remain in Canada for some time while his company dealt with a patent problem. When the problem was cleared up, he found himself fired from his company and betrayed by his business partner, Abraham White, whom he had considered "more than a brother." His first marriage ended around the same time. He left the American De Forest Company in the summer (Douglas 1987, p. 168; Lewis 1991, pp. 50–53). A year before, de Forest had been rich and famous; now he was broke and unemployed. He sought a job with Reginald Fessenden, whose lawsuit had just made him penniless, but in vain.

In his morally and financially difficult situation, de Forest concentrated on further experiments with the "audion."[36] In the summer of 1906 he took an important step, though he did not realized its full meaning at first: He wrapped the audion in a metallic sheath (figure 6.7) and applied for a patent for that modification. In the same patent, he also had the audion surrounded by the antenna. The precise reasoning behind these designs is not clear, but it appears that de Forest was trying to use the oscillation's electrostatic and electromagnetic effects to augment the audion's relaying effect. In figure 6.7, the metallic sheath, the gas, and the filament form a sort of condenser. The action of the outer sheath remained the same as before—that is, "the oscillating electric field developed by the electrical oscillations in the secondary circuit . . . operates to alter the conducting properties of the sensitive conducting gaseous medium in the vessel *D* and thereby to vary the current flowing in the local circuit"[37]—but de Forest believed that, owing to the "condenser arrangement," the signals received were "dull [and] muffled" rather than "sharp and staccato." Since dull and muffled signals were easily distinguished from static, de Forest (1906b, p. 747) found this device to be "most serviceable in practice."

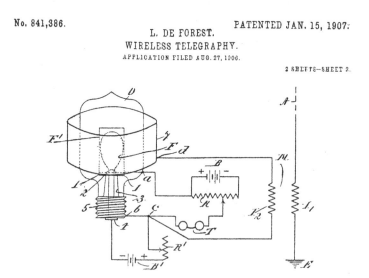

No. 841,386.   PATENTED JAN. 15, 1907.
                          L. DE FOREST.
                      WIRELESS TELEGRAPHY.
                      APPLICATION FILED AUG. 27, 1906.

                                              2 SHEETS—SHEET 2.

Figure 6.7
De Forest's audion with metallic sheath.

On October 26, 1906, de Forest read a paper titled "The Audion: A New Receiver for Wireless Telegraphy" at a meeting of the American Institute for Electrical Engineers. That paper was the public introduction of the work on two-electrode audions he had done before the summer of 1906. During the discussion, engineers asked de Forest a difficult question about the difference between the audion and Fleming's valve. Michael Pupin commented negatively on the paper, saying that he could not understand at all how the device worked ("Discussion," in de Forest 1906b, p. 764). Across the Atlantic, the abstract of de Forest's AIEE paper infuriated Fleming. By this time, however, de Forest had moved one step further. At the end of his AIEE paper (p. 762), de Forest mentioned his research on the "enclosed type of the Audion."

Just 24 hours before he read his paper, de Forest had filed a patent on several new audions in which magnetic effects were used to vary the conductivity of the sensitive gaseous medium. In two cases, the outer sheath of the previous audion was inserted inside the lamp (figure 6.8). That is, these audions now had two plates in the glass. One, connected to the antenna, had evolved from the outer sheath; the other was the standard plate linked to the positive pole of the local battery. The general principle remained

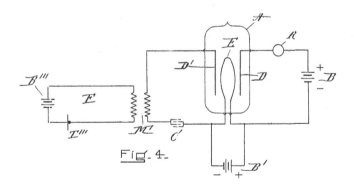

Figure 6.8
De Forest's audion with inserted plate (*D'*).

unchanged except for one important point: De Forest had begun to link his audions to signal amplification. For instance, he noted that *"the current to be amplified* may be impressed upon the medium intervening between the electrodes *D* and *E*, and thereby alter, by electrostatic attraction, the separation between the electrodes." The electrostatic action of the outer sheath (now inside the lamp) remained the same here, but signal amplification was a novel idea. Why de Forest emphasized amplification at this point, and whether his device actually amplified signals as he thought, is uncertain. As one commentator noted, however, "de Forest builded better than he (or anyone else of that day) knew."[38]

Late in November of 1906, de Forest hit upon the idea that "the third, or control, electrode could be located more efficiently between the plate and the filament" (de Forest 1950, p. 214). He first used a perforated plate for this, but soon he adopted a single zigzag grid of wire (figure 6.9). On November 25 he ordered several grid audions from the lamp maker H. W. McCandless & Co. A few days later, he was forced to officially resign from

No. 879,532.   PATENTED FEB. 18, 1908.
L. DE FOREST.
SPACE TELEGRAPHY.
APPLICATION FILED JAN. 29, 1907.

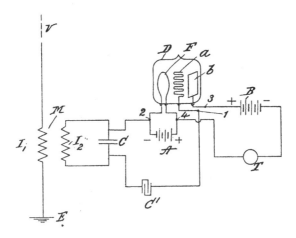

Figure 6.9
De Forest's audion with zigzag grid (a).

the American De Forest Company. In the margin of his letter of resignation he scribbled the following (Lewis 1991, p. 50–51):

This is the funeral of my first-born child! This the *finis* to the hopes and efforts which have made up my strenuous life for the past five years. That which I had wrought with pain and ceaseless endeavour to make grand and lasting & triumphant is prostituted, sandbagged, throttled & disabled by the Robber who has fattened off my brain. But my work goes on, while I live.[39]

De Forest's work did indeed go on. On December 1, 1906, he sent a telegram to his lawyer. When he met his lawyer the next day, he "sketched a rough diagram of the grid" on the back of a breakfast menu (Lubens 1942, January 24, p. 36). The patent application for the grid audion was written on December 23. The patent was filed on January 29, 1907. A few months later, de Forest organized the De Forest Radio Telephone Company, subsidized by the Radio Telephone Company, which sold grid audions and other radio equipment. Between 1907 and 1915, all the audions were manufactured by the McCandless Company. Between 200 and 600 were produced per year.[40]

De Forest's audion was not widely used. Relatively expensive ($5–$8), it was only a little more sensitive than simple crystal detectors. De Forest claimed that the audion amplified signals, but in fact it acted only as a rectifier, like Fleming's valve. In 1908 de Forest's own assistant called it "quite unreliable and entirely too complex to be properly handled by the usual wireless operator" (Chipman 1965, p. 99). Only amateur wireless enthusiasts were interested in purchasing audions (in order to tinker with them). As one engineer later recalled (Espenschid 1959, p. 1254), only a few amateur operators "managed to save up the required $5 and then suffered the filament burn-out soon to follow!"

The Birth of the Amplifying and the Oscillating Audion

An interesting similarity between the audion and the arc was that both produced a hissing or whistling sound. De Forest's first audion was born with a hiss. In a speech to the AIEE, de Forest (1906, p. 760) reported:

When the field is intense, a marked frying or hissing sound in the telephone is heard . . . In the hissing arc, parts of the arc are in rapid motion in the unstable portion around the edges of the positive terminal. Possibly also the presence of oxygen in the gas enters into the phenomena here as it does in those of the hissing arc. . . . In this particular Audion, I could get a great range of singing or squeaking sounds as the heating current was varied.

Such sounds were quite familiar to telephone engineers. The telephone set of the early twentieth century had a mouthpiece that was separate from the earpiece. When the former was brought close to the latter, the telephone produced a hissing or whistling sound. This, it was known, occurred because any tiny vibration entering the mouthpiece would be amplified by the relay or by other devices and would then come out of the earpiece, only to be fed back into the mouthpiece again as input. This cyclic process was repeated many times a second, creating the hissing or howling sound. The sound showed that the amplifying repeater was in fact producing some sort of sustained oscillation. In this way, feedback amplification was linked to the production of sustained oscillations.[41]

In 1911, in the laboratory of John Hays Hammond Jr. in Gloucester, Massachusetts, Fritz Lowenstein, an engineer with a background in power engineering, began work on a radio guidance system.[42] In order to control an object (for example, a boat or a torpedo) at a distance by means of electromagnetic waves, Lowenstein essentially needed three components: a good transmitter, a good amplifier for the receiver, and sensitive tuning. Since Lowenstein had worked as Nikola Tesla's assistant in power engineering and on the wireless transmission of power, he had some experience with the Hewett mercury lamp that was used both as a telephone repeater (amplifier) and as a converter between alternating and direct current in power engineering. He also knew that the mercury lamp, like the Duddell and Poulsen arcs, possessed a negative resistance. Because of this expertise, he paid attention to de Forest's audion, a device similar to the mercury lamp, and began to experiment with it to test whether it also possessed negative resistance, and, if it did, whether it could be employed for amplifying received signals like the mercury lamp in telephone engineering (Hammond and Purington 1967, p. 1197).[43] Lowenstein succeeded in designing (perhaps) the first audion amplifier circuit in November 1911. He and his assistants tested this circuit by connecting it to the telephone, and obtained clearer and much stronger telephonic signals in a long-distance experiment. Lowenstein wrote to Hammond: "When I heard your [Hammond's] voice I fairly jumped in delight; it came in so clear with every shade of its personal characteristic." (Hammond and Purington 1967, p. 1198) The amplifier did not sing or hiss (Miessner 1964, p. 19).

Lowenstein also designed an audio-frequency oscillator as a "steering circuit" for a torpedo system. In this steering circuit, an audion was connected to an *L-R-C* oscillatory circuit. Lowenstein's oscillatory circuit produced an audible high tone. He then developed it into a (relatively) high-frequency oscillating circuit. One day, Benjamin F. Miessner, who assisted Lowenstein in 1911–12, noticed that the audio-frequency sound from the steering circuit became faint and eventually disappeared when he decreased the capacitance of the circuit. Miessner and Lowenstein tested the circuit with a hot-wire ammeter to determine whether this phenomenon indicated that the audio-frequency sound was gradually changed into inaudible high-frequency oscillations, and found that it did. Lowenstein then experimented to see if he could use this circuit for wireless telephony, and he succeeded in transmitting human voices between two laboratories in the same building in early 1912. Without doubt this was "the first vacuum-tube continuous-wave radiotelephone transmission" (Hammond and Purington 1967, p. 1199). The frequency used was 15,000 Hz (Miessner 1964, p. 22).[44]

Lowenstein's amplifier did not attract the attention of American Telephone & Telegraph engineers. Lowenstein did not even apply for a patent for his oscillator, perhaps because he might have thought that "there was no invention, once the amplifying action of the tube was realized, especially since audions oscillated naturally" (Miessner 1964, p. 23). But his work did not go unnoticed. It came to the attention of Ernst Alexanderson of General Electric (the inventor of the Alexanderson high-frequency alternator), who informed Irving Langmuir. Lowenstein's work was also known to Beach Thompson, chairman of the Federal Telegraph Company in California. Thompson was in the midst of hiring de Forest (who had fled from New York to California) to perfect the Poulsen system. Having been informed of Lowenstein's use of de Forest's audion for amplification and oscillation, the Federal Telegraph Company asked de Forest to pursue this line of research and assigned him two assistants, Charles Logwood and Herbert van Etten. De Forest had begun to regain his health. One day in February 1912 he noted: "Each down brings hope; while a new health, an unknown physical strength, a renewed youth grows within me. It is California & I am only 38!" His notebook entry of April 22 records an experiment in which he used two rectifiers, or Fleming valves, in a wireless receiver.[45]

Although de Forest had been asked to develop an amplifier for the Federal Company, he had a different goal in mind. He was aware that AT&T was interested in securing a good amplifier for a proposed telephone communication between New York and California—an ambitious plan that AT&T had promised to accomplish by 1915.

In the summer of 1912, de Forest began a series of experiments on the use of the audion as an amplifier. One day, by connecting the output of one audion to the input of the other, he obtained good amplification (figure 6.10). At the same time, however, he heard a howling sound from the audion. If a telephone amplifier made a howling sound, this meant that it was useless as an amplifier. De Forest tried to eliminate the howling by changing circuit variables, but he found that he could not eliminate it completely. In October 1912 he took his device to New York to show it to John Stone Stone, an eminent engineer who had good connections with AT&T. When he demonstrated the amplifier to Stone, it still made the unwanted sound.[46]

De Forest's audion amplifier impressed Stone. It must also have impressed AT&T's engineers and managers, for AT&T soon bought exclusive rights to it for use in communication (except wireless communication).[47] But what did Lee de Forest actually invent in the summer of 1912? His notebook records the date of his first success in feedback amplification as August 6. This was the circuit in which audions were employed for the amplification of telephone signals. Does de Forest's invention of August 1912 include the amplification of radio-frequency oscillations? Does it include the generation of sustained oscillations? This is a subtle issue, because, as I mentioned earlier, although telephone engineers had known that amplification by repeaters was not separable from the production of sustained oscillations, whether the same was true for high-frequency oscillations was not obvious.[48] Moreover, de Forest tried to abolish, not maintain, the sustained oscillations that caused the unwanted howling sound. John Stone Stone later gave testimony in the patent litigation between de Forest and Edwin Howard Armstrong to the effect that in October 1912 he had asked de Forest whether he had known if the oscillations extended into the radio-frequency range, and that de Forest answered that he had known about it and had thought about using his circuit to generate such oscillations. Stone's testimony was crucial for establishing de Forest's priority over Armstrong in court. However, there is no other evidence to support Stone's claim. As several historians have noted, Stone was not free from AT&T corporate interest in de Forest's priority. At any rate, it is not historically meaningful to say that de Forest invented the vacuum-tube oscillator in the summer of 1912, because his aim was to avoid sustained oscillations of either low or high frequency.[49]

PLAINTIFF'S EXHIBIT No. 14 B

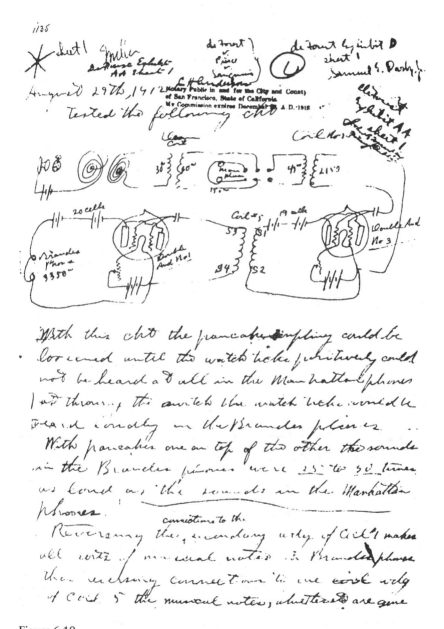

Figure 6.10
De Forest's feedback amplification circuit of August 29, 1912. Source: transcript of record, Supreme Court of the United States, October term, 1933 (no. 619), *Radio Corporation of America vs. Radio Engineering Laboratories, Inc.*

Armstrong consciously intended to utilize what de Forest wanted to eliminate. He was an amateur radio operator from his high school days.[50] In his early college years at Columbia University, around 1911, Armstrong obtained an audion from a friend who was an amateur radio-telegraphy operator. Though he connected the audion to the receiving circuit in many different ways, Armstrong could not get it to work as an amplifier. Armstrong thus developed his own theory of, and his own circuit for, the audion. Using an oscilloscope in the Columbia engineering laboratory, Armstrong measured the plate current (that is, the output) and the grid potential (that is, the input) of the audion. He found that the output fluctuated exactly as the input varied. What would happen, he wondered, if the output were somehow fed back into the input? It would be added to the input, and this increased input would once again increase the output. The same process would be repeated many times per second. This could yield a final output perhaps 100 times the original input. In September 1912, he confirmed his expectation: the audion "regenerated!" (See figure 6.11.) Armstrong later recalled this discovery as follows:

The invention was luck but the production of a workable apparatus was the work of a few hours—the unravelling of the phenomena involved in the system was a matter of months. Briefly the discovery [invention was crossed out] came out of a desire to find out how the audion worked—not an easy thing to do in the dark age of [19]11 and [19]12 when the very scanty literature on the subject spoke learnedly of "gas ionization" etc and the audion was known to the art simply as a detector of high frequency oscillations.[51]

With this amplifying circuit, Armstrong could clearly hear wireless signals coming from Ireland and even from Hawaii.[52]

On the same day that he discovered this feedback effect, Armstrong heard hissing from the audion receiver. While changing the inductance value, he found that clear wireless signals were replaced by a hissing note when the audion reached its maximum amplification point. This hissing then changed into a howling. Why did the circuit suddenly produce an audio-frequency sound? He soon theorized that the audion might in fact have produced high-frequency local oscillations, and that the howling was due to the beat produced by the superposition of these high-frequency local oscillations with incoming high-frequency signals. To test this hypothesis, Armstrong borrowed a sensitive ammeter and measured the plate current. Just beyond the maximum amplification point, he found that the current dropped suddenly. This was a typical sign of oscillation: the current dropped rapidly because the energy was transformed into electromagnetic radiation. The audion of 1912, one might say, was made to amplify and oscillate.[53]

Armstrong later maintained that he had created the amplification and oscillation effects of the audion in September 1912. The earliest evidence is a circuit diagram that he had notarized on January 13, 1913 (figure 6.11). Armstrong demonstrated his device to several people after January 13, and perhaps even before. With the notarized diagram, he was quickly able to establish priority over others, including the German Alexander Meissner (who filed a US patent in March 1913) and Irving Langmuir (who filed one on October 29, 1913).

Armstrong's patent application was filed the same day as Langmuir's. Armstrong had not filed for a patent earlier because he had been unable to raise enough money (about $200) to cover the patent fee and the attorney's fee. After filing, he told his attorney, William H. Davis, that he expected to extend the reception of wireless signals with his new detector "to Honolulu on the west and Italy and the northeast on the east." Early in January 1914,

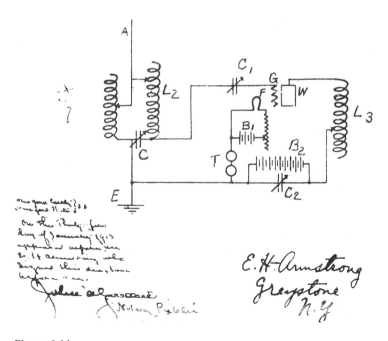

Figure 6.11
E. H. Armstrong's regeneration as drawn in January 1913. Source: transcript of record, Supreme Court of the United States, October term, 1933 (no. 619), *Radio Corporation of America vs. Radio Engineering Laboratories, Inc.*

Armstrong demonstrated his amplifying receiver to the American Marconi Company. The vice-president of the company reported to Marconi in Britain on the demonstration, noting that Armstrong was "a boy not much over twenty one." Armstrong's patent (US Patent 1,113,149) was issued on October 6, 1914.[54]

De Forest filed a patent on the oscillating audion (the "ultra-audion," as he called it) on March 20, 1914, and a broader patent on the feedback circuit on September 25, 1915. In the latter he stated that he had invented the feedback circuit before March 1913. Armstrong sued de Forest for infringement of his patent. De Forest defended his patent by asserting that his observation of the howling sound in August 1912 was the same as Armstrong's invention in September 1912. The judge in New York District Court, Julius Mayer, ruled in favor of Armstrong on the basis of two pieces of evidence: (1) If de Forest had known the true meaning of his invention, he would have applied for a patent for it quickly, as he did for several patents during that period. (2) De Forest's 1914 notebook on the "ultra-audion" shows that even at that time de Forest did not fully understand the principles involved in Armstrong's feedback circuit. The notebook shows that de Forest proceeded from complete ignorance to ineffective patented circuits. The judge ruled that de Forest's claim to have known about the full-fledged feedback circuit before March 1913 was not convincing.[55]

However, there was another dispute going on: the "four-party" interference proceedings[56] involving Armstrong, de Forest, Langmuir, and Meissner. At the interference proceedings, the examiners decided unanimously in favor of Armstrong. However, de Forest appealed to the Court of Appeals of the District of Columbia, and the decision was reversed. De Forest's experiments in the summer of 1912 were found to constitute an invention of the "means for producing sustained electrical oscillations." After this success, de Forest's was granted patents on the ultra-audion and the regenerative circuit. He then sued Armstrong in Pennsylvania for infringement of those patents. The court decided in favor of de Forest. Armstrong appealed, but the Appeals Court affirmed the lower courts' decision. The Supreme Court denied a petition. The result of this lawsuit was an agreement on a decree that invalidated most of Armstrong's patent claims.

This was not the end of the story. In 1934, AT&T, which owned de Forest's patent, sued a small manufacturing company for infringement.

Armstrong decided to pay the company's litigation expenses. The District Court decided that de Forest's patent was invalid and admitted Armstrong's priority. An Appeals Court reversed the decision. The US Supreme Court ruled for de Forest. After this defeat, Armstrong returned the Institute of Radio Engineers' 1917 Medal of Honor for the feedback circuit, but the IRE's board of directors unanimously reaffirmed their original decision (Gannett 1998).

Epilogue: The Making of the Radio Age

The audion revolution alone did not create radio broadcasting. Social and cultural factors—including amateur operators, World War I, corporate competition, and other socio-cultural issues—should not be underestimated (Douglas 1987). In this book I have focused on one important strand of the story: the complicated trajectories of several technologies and their crossovers. I have shown how engineers and scientists gradually, and sometimes haphazardly, exploited various scientific effects and technological artifacts, with which they eventually produced and received continuous waves. Although technology alone does not make society, technology opens new possibilities and closes some old ones. Whenever this happens, people are forced to think, choose, exploit, and adapt to these new possibilities. Different groups of people have different interests in technologies, and their different preferences and choices often create cultural tensions and social conflicts.

In the 1910s the amplifying and oscillating audion made the production, transmission, and reception of continuous waves much easier and cheaper. Before the audion revolution, it was expensive to produce continuous waves. Gigantic high-frequency alternators cost a million dollars; reliable arc generators were also bulky and expensive. These were power technologies, and only big corporations that could afford these transmitters were sending continuous waves carrying human voices. After the audion revolution, it became much easier for anyone to set up a small transmitting station. Power mattered little, but interference became a central issue as the ether became busier. Starting in the early 1920s, many radio broadcasting stations were built. The allocation of the spectrum became the subject of intense public debate. Other important technological innovations—such as

Armstrong's superheterodyne circuit, the superregenerative circuit, and frequency modulation (FM)—followed the oscillating audion.[1] In early wireless telegraphy, Marconi thought in terms of inductance, capacitance, and
resistance—all variables in cable telegraphy. Starting in the 1920s, radio
engineers thought in terms of the audion and its associated effects. The
audion redefined theory and practice in radio engineering. It may not have
been the sole creator of radio broadcasting; but it certainly did open new
possibilities and agitate existing ones by making the spectrum cheap and
easily accessible.

Appendix: Electron Theory and the "Good Earth" in Wireless Telegraphy

Most literature on transatlantic wireless telegraphy, including Marconi's various recollections, has emphasized that some eminent scientists, including Lord Rayleigh and Henri Poincaré, had objected to Marconi's plan by arguing that electromagnetic waves propagate only linearly. It has been also emphasized that Marconi was never discouraged by such merely "theoretical" objections, because he had a firm belief, based on practice, in the transmission of electromagnetic waves across oceans. However, the story is not strictly true in two senses. First, it was after Marconi's success in December 1901 that the debate on the transmission of electromagnetic waves occurred among Lord Rayleigh, H. M. MacDonald, Poincaré, and others. Second, there was little objection to Marconi's transatlantic wireless telegraphy simply because Marconi's experiment was not widely known to the public before December 1901. Further, as we have seen, many eminent engineers and scientists presumed that electromagnetic waves could travel below the curved surface of the earth in some way or another. Here I will detail Fleming's conception of the creeping character of the electromagnetic waves generated from Marconi's transmitter.

Fleming's conception was based on J. J. Thomson's and Joseph Larmor's electron theory, which Fleming adopted around 1900. In 1902, Fleming gave a Friday Lecture at the Royal Institution on the "Electronic Theory of Electricity." Although Fleming knew that the electronic theory of electricity was an extension of Weber's idea rather than Maxwell's, he also considered that it did not violate, but rather supplemented, Maxwell's theory, because the electron was a strain center in the ether, or a locality from which ether strain radiated. Therefore, the ether's movement was intimately related to the electron's movement. It was because "an electron in motion is in fact a shifting center of ether just as a *kink or knot* in a rope can be changed from place to place on the rope" (Fleming 1902, p. 177). The influence was reciprocal. Radiation, for example, was generated by the accelerating motion of electrons; electrons were affected by such a motion of the ether as radiation because they were the centers or converging points of the strain of ether.

The amalgamation of the Larmorian electron theory with practical wireless telegraphy appeared in Fleming's 1903 Cantor Lecture (Fleming 1903a). According to the electron theory, production of the ether strain occurs as illustrated in figure A.1. Two metal rods are placed in line and separated by a small spark gap. If the motion of electron is sufficiently rapid, the lines of strain in the external medium cannot contract or collapse quickly enough to keep up with the motions of the electrons. This motion of the self-closed lines of electric strain, then, constitutes electric radiation (ibid., p. 715).

This type of radiator was, however, Hertz's, not Marconi's. Marconi's radiator consisted of a long vertical wire, the lower end of which was attached to one of the two spark balls near the earth, the other spark ball being connected to the earth. Fleming (ibid., p. 744) mentioned that "it was the introduction of the aerial or radiator [antenna] by Mr. Marconi which laid the foundation for Hertzian wave telegraphy as opposed to mere experiments with the Hertzian waves." The critical difference that Fleming mentioned can best be understood by considering the production of electrical strain from Marconi's radiator in terms of the electron theory (figure A.2). Because the lines of the electric strain should terminate on an electron or a co-electron,[1] these semi-loop lines march outward with their feet on the ground. The same electrons do not always need to travel along the earth: "Since the earth is a good conductor," said Fleming (1903a, p. 718), "we must suppose that there is a continual migration of the electrons forming the atoms of the earth, and that when one electron enters an atom, another leaves it."[2] The earth, as a result, became a huge "wave-guide."[3]

Some interesting consequences followed from this reasoning. First, it could explain the dependence of the Hertzian telegraphy on the nature of surface over which it is conducted and on the atmospheric conditions. Experience had showed that sea was much better than land in carrying the Hertzian wave farther, and that night was better than day. Fleming thought that sand or dry earth would be less effective than wet earth.[4] Second, and

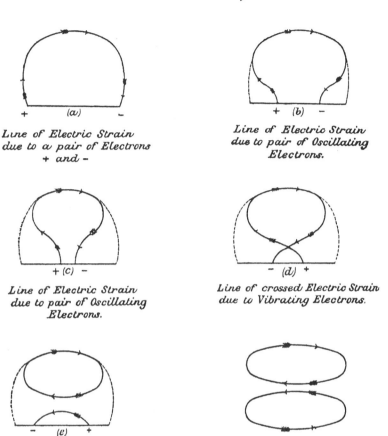

Figure A.1
Successive stages in the generation of electric strain from a Hertzian oscillator with a spark gap in its middle. Source: Fleming 1903a, p. 715.

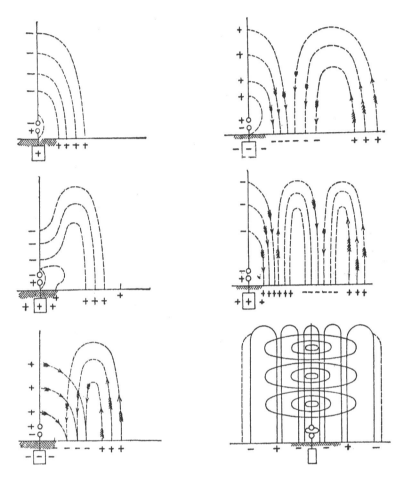

Figure A.2
Successive stages in the generation of electric strain from Marconi's radiator with one pole grounded. Source: Fleming 1903a, p. 719.

more important, the earth as a wave-guide implied the possibility of long-distance wireless telegraphy. Fleming's belief that the earth would guide the Hertzian wave stimulated him to continue his researches on long-distance wireless telegraphy. In his 1903 lecture, Fleming proposed diffraction or bending of electric wave around the earth as a cause of transatlantic transmission of wireless signals. But he also asserted that "Hertzian wave telegraphy is not that simply of a free wave in space, but the transmission of a semi-loop of electric strain with its feet tethered to the earth, it is possible that . . . an ether disturbance could be made in England sufficiently powerful to be felt in New Zealand" (Fleming 1903a, p. 781). In this sense, a "good earth," as in old telegraphy, was essential for long-distance Hertzian telegraphy.

Notes

Chapter 1

1. See, e.g., Rowlands and Wilson 1994; Burns 1994.

2. For Süsskind's work on Popov, see Süsskind 1962. On the 1995 controversy, see Aldridge 1995; Vendik 1995; Barrett and Godley 1995.

3. Since my text focuses on scientists and engineers who were educated in or who worked on Hertzian waves in Britain, here I will briefly examine here the case for scientists and engineers in countries other than Britain. First, it is widely agreed that the research of Augusto Righi in Italy and Jagadis Bose in India on "microwave optics" had nothing to do with wireless telegraphy. However, it is worth mentioning that Righi had some impact upon Marconi, and Bose influenced Henry Jackson to work on wireless signaling. For Righi, see Aitken 1976, pp. 183–185; for Bose, see Dasgupta 1995–96. In Russia, A. S. Popov, inspired by Lodge's 1894 lecture, conceived of the possibility of employing Hertzian waves for "transmission of signal over a distance" in 1895. As Süsskind (1962) shows, there is indirect evidence that Popov secretly demonstrated signal transmission over a short distance in March 1896, but at that time Marconi had already constructed and demonstrated half-practical wireless telegraphy. In the US, Tesla undoubtedly thought about possible ways of using rapid oscillation for signaling in the early 1890s. Tesla's admirers claim that Tesla demonstrated message transmission by means of Hertzian waves for the first time in 1893 in his St. Louis lecture. This claim is, however, not supported by any direct evidence, and the fact that Tesla used a Geissler tube as a detector (how could one effectively detect Morse-coded signals with the Geissler tube?) strongly weakens the claim. In addition, his ideas about how signals are communicated through space were similar to earth-conductive (wireless) telephony, rather than Hertzian wireless telegraphy. For the Tesla-first claim, see Cheney 1981, pp. 68–69. Compare Cheney's claim with Tesla's original, unclear ideas (Tesla 1894, esp. pp. 346–349). See also Anderson 1980. To my knowledge, no German scientists or engineers conceived of the possibility of Hertzian wave telegraphy before Marconi. Both Adolf Slaby and Ferdinand Braun initiated research on Hertzian wave telegraphy after they heard the news of Marconi's practical success (Slaby 1898; Kurylo and Süsskind 1981).

4. On Lodge's experiments on the air gaps of lightning conductors, see Lodge 1890b, pp. 352–353. For the story of Branly's tube in Britain, see Lodge 1897. See also Branly 1891 and Phillips 1980, pp. 18–37.

5. Minchin's experiment is reconstructed from the recollection of Rollo Appleyard, who had assisted Minchin; see Appleyard 1897. Lodge's experiment at Liverpool is quoted from Rowlands 1990, pp. 116–117.

6. On FitzGerald's formula, see FitzGerald 1833. See also Hunt 1991, p. 43. For Lodge, see Lodge 1894b, pp. 135–137. See also Lodge's recollection (Lodge 1931, p. 165).

7. Lodge remarked that William Thomson first used the term "electric eye." See Lodge 1890a.

8. Trotter had worked under Maxwell at the Cavendish Laboratory of Cambridge University for a short time in the late 1870s, and served the editor of *The Electrician* since the late 1880s. See [A. P. Trotter], "Notes," *The Electrician* 26 (April 10, 1891), p. 685.

9. "Editorial," *The Electrician* 39 (24 Sep. 1897), 699. How Threlfall's comment was brought to light is interesting and worthy of mention. It had not been noticed at all until Oliver Lodge informed J. J. Fahie of it in 1899. Fahie mentioned Threlfall's lecture in a footnote in the book he was writing on the history of wireless telegraphy. Subsequently, Threlfall was considered to be one of those who preceded Marconi. (See Fahie 1899, p. 197, note 2.) On the rehabilitation of Threlfall by modern historians, see Süsskind 1969a, pp. 69–70.

10. On the life and work of Crookes, see D'Albe 1923.

11. Hughes's experiment came into light in 1899, when J. J. Fahie noticed this anachronistic passage in Crookes's article. Fahie asked Crookes about this, who replied that this experiment was done by Hughes with a microphone in 1879. Fahie then wrote Hughes and obtained the first detailed description of Hughes's experiments. Crookes's reply and Hughes's letter were published as appendix D of Fahie 1899, accompanied by the statement that "Hughes's experiments of 1879 were virtually a discovery of Hertzian waves before Hertz, of the coherer before Branly, and of wireless telegraphy before Marconi and others." For a more balanced description of Hughes's experiments, see Süsskind 1968a, pp. 97–98.

12. R. E. B. Crompton mentioned this possibility; see Lodge 1923, p. 332. On Willoughby Smith's induction telegraphy, see Smith 1888 and Fahie 1899, pp. 162–166. To the public, Hertzian wave telegraphy and induction wave telegraphy were not discernible. When William H. Preece performed an experiment on induction telegraphy, the Victorian magazine *Spectator* described Preece's experiment as the one that had been predicted by Crookes; see "A Dreamy View of Mr. Preece's Experiment," *Spectator* 19 (1892), November 26: 764–765.

13. See Crookes 1891b and "Science and Conjecture," *Spectator* 67 (1891), November 21: 723–724.

14. See "Notes," *The Electrician* 28 (February 5, 1892), 341–342.

15. J. A. Fleming, "A Few Notes on the Method of Presenting the Marconi Case" (April 26, 1904), MS, Marconi Company Archives (henceforth MCA).

16. "Wireless Telegraphy," *Electrician* 39 (1 Oct. 1897), 736.

17. See, e.g., the letter to Lodge in which Minchin states: "If you think that anything that I have can be combined in any way with your arrangements so as to solve the problem of wireless telegraphy, I shall be happy to join. . . . Does Marconi want to prevent us from earthing our poles? I gather that he does." (Minchin to Lodge, November 1, 1897, Lodge Collection, University College London) Marconi seem to sense this: "I hear, however, that Lodge is going to make a union or league with all other holders of Patents connected with Wireless Telegraphy in order to fight me and my Company." (Marconi to Preece, November 16, 1897, MCA)

18. The story about Rutherford's research is based on E. Rutherford, Laboratory Notebook, Rutherford Papers MS Add 7653 NB 2, pp. 6–80, Cambridge University Library, Cambridge. See also D. Wilson 1983, pp. 87–89.

19. For Rutherford's letter, see Eve 1939, pp. 22–33; see also D. Wilson 1983, pp. 95–96. For Thomson's letter to Kelvin, see J. J. Thomson to Kelvin (undated, but written around February 27, 1896); J. J. Thomson to Kelvin (April 10, 1896), Kelvin Collection, Add 7342, Cambridge University Library. Rutherford's work was published as "A Magnetic Detector of Electrical Waves and some of its Applications" (Rutherford 1897). For Rutherford's recollection of his research on Hertzian waves, see Rutherford 1902. Rutherford's presentation before the Cambridge Natural Science Club is recorded in Cambridge University Natural Science Club: Minute Book, MS, Science Periodical Library, Cambridge, p. 554.

20. On Jackson's research, see Jolly 1972, pp. 71–72. For an example of an exaggerated historical account of Jackson's contribution, see Pocock 1963.

21. For a description of Bose's coherer, see Bose 1895b. For *The Electrician*'s comment on it, see "Notes," *The Electrician* 36 (1895), December 27: 273. For Bose's work on "microwave" optics, see Bose 1895a. In Calcutta, Bose demonstrated before the public the production and reception of Hetzian waves by ringing a bell and by firing a gun in 1895. On Bose's life and work, see Dasgupta 1995–96.

22. "Statement of Cap. Jackson's Claim as regards the invention of Wireless Telegraphy," in "Report of Cap. Hamilton" (January 28, 1899), ADM 116/ 523, Public Record Office, Kew.

23. "Report of Captain Jackson," (16 Sep. 1896) in ADM. 116/ 523, Public Record Office, Kew.

24. "Report of Captain Jackson," May 22, 1897, ADM 116/523, Public Record Office, Kew.

25. Marconi's recollection of when he took lessons from Professor Rosa is in conflict with that of Professor Giotto Bizzarrini, who also taught Marconi in Leghorn. Marconi recalled that he met Rosa before he was eighteen (i.e., before 1892). However, in Bizzarrini's recollection, Marconi's brother, Alfonso, introduced Guglielmo to him in 1892, and Marconi's mother introduced Guglielmo to Professor Rosa in the same year. According to Bizzarrini, the 18-year-old Marconi was interested in chemistry, the Ruhmkorff spark coil, and telegraphic apparatuses, and he did some research on electrical oscillations caused by atmospheric discharges. See Bettòlo 1986.

26. Marconi later emphasized that he learned physics from Professor Rosa, but did not mention his indebtedness to Righi. See Marconi 1909, p. 196. Righi, however, recalled that he had appreciated Marconi's "inventive power and his unusual intellectual gifts" when he first saw Marconi; see *Nature* 66 (9 Oct. 1902), 581.

27. Marconi's research was reconstructed from his testimony in *Marconi Wireless Telegraph Company of America vs. De Forest Wireless Telegraph Company: Complainants Record for Final Hearing* (United States Circuit Court, 1904), pp. 530–540. This is his most detailed and systematic recollection on his research in Italy. I think this testimony is very credible, because Marconi's description of his receiver here is quite compatible with what other people reported on his receiver in 1896 and 1897. See H. R. Kempe, "Signalling Across Space by Marconi System" (a report to W. H. Preece), 15 Sep. 1896, Marconi Company Archives; G. S. Kemp, "Diary of Wireless Expts at G.P.O. & Salisbury," Marconi Company Archives; [W. H. Preece], "Report on Recent Experiments with the so-called Wireless Telegraphy," 29 Oct. 1897, Marconi Company Archives. Marconi also provided a detailed description of his early receiver in his first engineering paper, Marconi 1899.

28. On p. 536 of *Marconi Wireless Telegraph Company of America vs. De Forest Wireless Telegraph Company*, Marconi describes the high inductance as "insulators for Hertzian waves of high-frequency oscillation."

29. Marconi said that he took this from his notes. See Marconi to Preece, November 10, 1896, MCA HIS62. Marconi's method contrasts sharply with Oliver Lodge's. In the early 1890s, Lodge had connected one end of the receiver to the gas pipes of his laboratory, but he found this to be disadvantageous (Lodge 1932, p. 233). Connecting one end of the transmitter to the pipes should have been avoided, since it increased the wavelength by doubling the capacitance.

30. Marconi's grounded transmitter distinguishes his antenna from other similar inventions. Popov used a coherer and a long wire in his lightning detecting device, but his inspiration seems to have originated more from a lightning rod than a telegraph. Minchin used a long wire for his receiver (which he called the "feeler") and Jackson also attached wires to the coherer (which he called "wings"), but none of them grounded the transmitter. Pocock's (1988, p. 100) comment that "Marconi had, like Popov and Jackson, added aerials to his apparatus" does not fully recognize the originality in Marconi's antenna design.

31. Preece's lack of interest in Hertzian waves was pointed out in "The Transmission of Electric Signals through Space," *The Electrician* 31 (1893), September 15: 520–521.

32. See, e.g., the recollection of the Australian engineer George W. Selby (1898) of his early attempt to devise a practical wireless system.

33. For more on this, see chapter 3 and the appendix.

Chapter 2

1. For example, because of Lodge's 1894 demonstrations, the historian of technology Hugh Aitken (Aitken 1976, p. 123) has argued that wireless telegraphy cannot

be said to have been invented by Guglielmo Marconi: "Did Lodge in 1894 suggest in public that his equipment could be used for signaling? Did his lecture refer to the application of Hertzian waves to telegraphy? Did he demonstrate transmission and reception of Morse Code? The answer would seem to be affirmative in each case. In this sense Lodge must be recognized as the inventor of radio telegraphy." Stranges (1986), reviewing the second edition of Aitken 1976, noticed this point.

2. For claims supportive of Marconi's priority, see Süsskind 1962, 1969a,b. Aitken's claim for Lodge's priority was not unprecedented. W. P. Jolly, who has written biographies of both Lodge and Marconi, admitted Lodge's wireless telegraphy at the British Association meeting in 1894 (Jolly 1972, pp. 41–42; Jolly 1974, p. 97). After Aitken, however, Lodge's priority was widely accepted. A recent biography of Lodge emphasizes Lodge's "radio transmission" in 1894, based upon Aitken's account and Lodge's own. Rowlands 1990, pp. 115–123. Pocock (1988), though admitting Marconi's originality, mentions Lodge's radio-transmission in the Oxford lecture in 1894, on p. 82. G. A. Isted, a former assistant to Marconi, has lately written that Lodge's demonstration at the British Association in Oxford "is the earliest recorded instance of the transmission and reception of a signal by Hertzian waves and it is clearly of great historical importance" (Isted 1991a, p. 46). Aitken's argument was also picked up by Basalla (1988, p. 99).

3. On the lives and works of the British Maxwellian physicists, see Buchwald 1985 and Hunt 1991.

4. The use of patent records and patent interferences as sources for historical research has been pointed out by Reingold (1959–60) and by Chapin (1971). See also Hounshell 1975; Post 1976; Brittain 1970.

5. "Dr. Oliver Lodge's Apparatus for Wireless Telegraphy," *The Electrician* 39 (1897): 686–687. Also quoted in Aitken 1976, p. 122.

6. The British Association's annual meeting was held at Oxford in August 1894. The date of Lodge's lecture and experiments was August 14.

7. See Fleming 1937, p. 42.

8. John Ambrose Fleming to Oliver Lodge, August 24, 1937, Lodge Collection, University College London.

9. (Copy of) Lodge to Fleming, August 26, 1937, Lodge Collection, UCL.

10. Fleming to Lodge, August 29, 1937, Lodge Collection, UCL.

11. Ibid.

12. On Lodge's early conceptions on electromagnetic waves, see Buchwald 1994 and Hunt 1991, pp. 24–47. For the best description of Lodge's research after 1888, see Aitken 1976, pp. 80–102. For Lodge's promulgation of his program with Hertzian waves and his concept of "imperial science," see Lodge 1889, pp. 303–307. For the early quasi-optical experiments with Hertzian waves, see Ramsay 1958.

13. The distance was about 70 yards.

14. The lecture, "The Work of Hertz," was published in *Nature*, in *The Electrician* (with illustrations), and later in *Proceedings of the Royal Institution*. The reference

here is Lodge 1894b. The lecture, with some appendixes, was published in 1894 as a book (Lodge 1894c). With the third edition (1900), its title was changed to *Signalling through Space without Wires.*

15. The bell was not connected to the coherer circuit, nor did it tap the coherer directly.

16. "Hertzian Waves at the Royal Institution," *The Electrician* 33 (1894): 156–157.

17. A "dead-beat" galvanometer used a needle with a small moment of inertia. If the current changed suddenly, the needle moved from one point to another, where it stopped "dead."

18. Lodge (1894b, p. 137) mentioned "the spot of light" in mirror galvanometer.

19. Lodge's exhibited the portable detector of his assistant's design at the Royal Society soirée a few days after his Friday Lecture. See "The Royal Society Conversazione," *Nature* 50 (1894): 182–183.

20. Oliver Lodge to J. Arthur Hill, December 11, 1914, in Hill 1932, p. 47.

21. For the same recollection, see Lodge 1931, p. 164, and Lodge 1932, p. 231.

22. Muirhead was so excited after Lodge's Oxford lecture that "the next day he went to Lodge with the suggestion that messages could be sent by use of these waves to feed cables" (Muirhead 1926, p. 39, quoted in Pocock 1988, p. 83).

23. Thomson's marine galvanometer was a very sensitive current-measuring device specially designed in such a manner that the swing of a ship could not change the readings. In principle, it utilized rotation of a small magnet fixed in the middle of the coils by silk fiber. When magnetic fields were created around the coils by the action of current, the small magnet was forced to rotate, and this effect was magnified by the reflection of a ray of light from a small mirror fastened to the magnet. For a detailed description of the device, see Prescott 1888, pp. 154–157.

24. For instance, Lodge (1889, p. 300) used the Thomson marine galvanometer lent by Muirhead for his experiments on electric momentum. Notice also that their business relationship began around the time that the Muirhead Company began to construct Lodge's lightning guard (Lodge 1892, pp. 419–426). I thank Ido Yavets for the latter reference.

25. Oliver Lodge, "Notes on the History of the Coherer Method of Detecting Hertzian Waves and other Similar Matters" (n.d.), Lodge Collection, UCL. In the published paper (Lodge 1897, p. 90), a similar paragraph read as follows: "Almost any filing tube could detect signals from a distance of 60 yards, with a mere six-inch sphere as emitter and without the slightest trouble, but the single-point coherer was usually much more sensitive than any filing tube."

26. Since the lectures were not published, I rely on the brief reports of the meetings of the British Association published in *Nature*, in *The Electrician*, *Engineering*, and in the London *Times*, all of which sent reporters to the British Association. "Physics at the British Association," *Nature* 50 (1894): 408; "The British Association at Oxford: Tuesday, August 14th," *The Electrician* 33 (1894): 458–459; "The British Association, Section A: Electric Theory of Vision," *Engineering* 58 (1894): 382–383; *Times*, August 15, 1894. (N.B.: Here and below, "*Times*" refers to the London newspaper.)

27. "The British Association at Oxford: Tuesday, August 14th," *The Electrician* 33 (1894): 458; "The British Association, Section A: Electric Theory of Vision," *Engineering* 58 (1894): 382–383; *Times*, August 15, 1894.

28. "Dr. Oliver Lodge's Apparatus for Wireless Telegraphy," *The Electrician* 37 (1897), p. 686.

29. On this controversy, see Hunt 1983; Jordan 1982; Yavets 1993.

30. For the description of Marconi as a practician, see "Notes," *The Electrician* 39 (1897): 207.

31. Different opinions have existed about the relation between Preece and Marconi. Aitken (1976, pp. 210–216) suggests that Preece's interest came from the "bureaucratic responsibility" of Preece and the Post Office to oversee the development of all forms of electric communication in Britain. Based upon the manuscripts of the Post Office, Pocock (1988, pp. 114–117) shows that Preece was rather cool toward the Marconi system's commercial possibility, then argues that Preece in fact followed the Post Office's policy with respect to new inventions—"neither to accept the invention, nor to invest substantial sums"—without entirely ignoring Marconi's new invention. But Pocock seems to have difficulty in explaining why Preece ardently advertised Marconi in the British Association and in his public lectures. As Nahin (1988, p. 281) mentions, the difficulty disappears if the personal factors are taken into account.

32. See also "Physics at the British Association," *Nature* 54 (1896): 567; *Times*, September 23, 1896; "Notes," *The Electrician* 37 (1896): 685. Preece also mentioned Marconi's parabolic antenna in the transmitter, and a relay and a Morse inker in the receiver.

33. George F. FitzGerald to Oliver Heaviside, September 28,1896, Heaviside Collection, Institution of Electrical Engineers, London.

34. Lodge to Fleming, August 26, 1937. Lodge's remark on Preece is in Lodge to Hill, December 11, 1914.

35. See also Pyatt 1983, pp. 12–35.

36. Oliver Lodge to Silvanus P. Thompson, 16 Mar 1897, Lodge Collection, UCL.

37. When asked about the difference, Marconi answered: "I don't know. I am not a scientist, but I doubt if any scientist can tell you." (Dam 1897) For an example of how much the "Marconi wave" upset Silvanus Thompson, see J. S. Thompson and H. G. Thompson 1920, p. 81.

38. Concerning Marconi's secret box, there is an interesting story. When Frederick Trouton, an assistant of FitzGerald, found an ordinary glass-tube coherer in Marconi's secret box, Marconi slammed it down again, saying "You would steal my invention." On this, Jolly 1974, p. 148. FitzGerald seems to have first solved the puzzle of Marconi system. He analyzed that "what Marconi is doing with his kites, poles &c &c, is to manufacture an enormous radiator and it is not the short waves of his double ball arrangement that he is emitting and receiving but the very much longer waves of his whole system. By connecting to earth he uses the earth as the second plate of his transmitter. . . . Anyway a *big* open system is the thing." (George F. FitzGerald to Oliver Lodge, October 30, 1897, Lodge Collection, UCL)

206 Notes to pp. 39–43

39. "The Man in the Street of Science," *The Electrician* 39 (1897): 546–547.

40. Oliver Lodge to Silvanus P. Thompson, June 1, 1897, Lodge Collection, UCL.

41. "Notes," *The Electrician* 39 (1897): 207.

42. George F. FitzGerald to Oliver Lodge, June 21, 1897, Lodge Collection, UCL.

43. Oliver Lodge, "Telegraphy without Wires," *Times*, June 22, 1897.

44. Lodge to Thompson, June 1, 1897.

45. FitzGerald to Lodge, June 21, 1897.

46. "Notes," *The Electrician* 39 (1897): 237. See also *Nature* 56 (1897): 185.

47. Oliver Lodge, "Improvements in Syntonized Telegraphy without Line Wires," 11,575, Provisional Specification (application May 10, 1897; complete specification February 5, 1898; acceptance August 10, 1898). For Lodge's syntony, see Aitken 1976, pp. 130–142.

48. Aitken 1976, pp. 285–286, note 12. Guglielmo Marconi, "Improvements in Transmitting Electrical Impulses and Signals, and in Apparatus Therefor," 12,039, Provisional Specification (application June 2, 1896). The content of the patent, of course, had been kept secret until its complete specification was accepted on July 2, 1897.

49. Even Lodge admitted Marconi's novelty in the tapping system. See, Oliver Lodge, "Report to the Chief Engineer of the Government Telegraphs" (June 1900), in ADM. 116. 570, Public Record Office, p. 5. In the same document, Lodge concluded (on p. 25) that "what Marconi can righteously claim is that he has made the coherer work dependably and give good signals in ordinary Morse code, and that he has extended the method overemarkably great distances: by the employment of adequate power, and by the appropriate means of an elevated and earthed wire like a lightning conductor."

50. For a contemporary witness's comments on Marconi's antenna, see Slaby 1898, esp. pp. 870–871. Even Lodge (1900, p. 47) admitted that Marconi's antenna was highly original.

51. FitzGerald to Lodge, June 21, 1897.

52. "Notes," *The Electrician* 39 (1897): 431.

53. John Fletcher Moulton (1844–1921) was the first Smith's Prizeman and Senior Wrangler of the Mathematical Tripos in Cambridge, in 1868. He soon became Fellow of the Royal Society due to his electrical research, and then engaged in legal works. See Moulton 1922; "John Fletcher Moulton," *Dictionary of National Biography* (1912–1921), pp. 392–394.

54. Silvanus P. Thompson to Oliver Lodge, June 30, 1897, Lodge Collection, UCL.

55. It was on July 20, 1897, and the name of the company was "The Wireless Telegraph and Signal Company." In February 1900, the name was changed into "Marconi's Wireless Telegraph Company." For the early history of the Company, see W. J. Baker 1970, pp. 35 ff.

56. Under the British patent system at that time, in which the comptroller of the Patent Office had no power over the contents of the patent, an inventor could claim

as many inventions as he wanted in a single specification at his own risk. In cases of some new inventions, an inventor could deliberately forge the claims with the effect of monopolizing the "principle" of that invention, rather than merely a specific artifact. Marconi's patent was close to such cases. James Watt's powerful patent on his new steam-engine with a separate condenser is another example. See "Patent," *Encyclopedia Britannica* (Chicago, 1972), p. 451. (The "Patent" entry in the most recent *Britannica* is not helpful for this.) See also Robinson 1972.

57. Guglielmo Marconi, "Improvements in Transmitting Electrical Impulses and Signals, and in Apparatus Therefor," 12039 in 1896, Complete Specification. The patent is also printed in Fahie 1899, pp. 296–320.

58. "Notes," *The Electrician* 39 (1897): 665. For FitzGerald's comment, see FitzGerald to Lodge, October 30, 1897.

59. Just after Marconi's patent was published, *The Electrician* published a series of articles on the coherer (including Lodge 1897).

60. For this episode, see *Times*, August 15, 1894); Lodge 1931, pp. 162–163. Even after this, Lodge often used the term "Maxwellian wave"; see, e.g., Lodge 1897, p. 89.

61. Silvanus P. Thompson, "Report of Wireless Telegraph Patents," (1900) ADM. 116. 570, Public Record Office, p. 38.

62. "Notes," *The Electrician* 39 (1897): 665.

63. "Dr. Oliver Lodge's Apparatus for Wireless Telegraphy," *The Electrician* 39 (1897): 686–687.

64. Lodge, "Report to the Chief Engineer of the Government Telegraphs."

65. John Ambrose Fleming criticized Lodge's book as "a perversion of fact" (Fleming to Guglielmo Marconi, January 12, 1900, Marconi Company Archives).

66. Oliver Lodge to William H. Preece, 4 Mar 1898, in E. C. Baker 1976, pp. 299–300. On Lodge's and Preece's magnetic induction telegraphy, see Lodge 1898b and Preece 1898.

67. Lodge, "Report of the Chief Engineer of the Government Telegraphs"; Thompson, "Report of Wireless Telegraph Patents." Thompson (p. 36) summarized Marconi's claims as either invalid or unessential.

68. For an account of this corporative politics around 1900, see Jolly 1972, pp. 68 ff. See also Cap. C. G. Robinson, "Synopsis of Report," (26 Sep. 1900) ADM. 116. 570, Public Office, where Cap. Robinson suggested a strategy of nullifying Marconi's patent by claiming Cap. Jackson's priority. Cap. Jackson's Report on Lodge's and Thompson's Reports is in the form of a letter, Cap. Jackson to J. A. Fisher, in ADM. 116. 570, Public Record Office, where Henry Jackson insisted that "a Governmental Office would probably fail in their suits, if they were to contest the validity of the Marconi Patents," and that "it would be a better policy to enter into negotiations with the company . . . than to enter into costly litigation with them."

69. The Morse Code signal for the letter S is three dots.

70. Oliver Lodge to Silvanus P. Thompson, April 11, 1902, Lodge Collection, UCL.

71. Fleming, "Wireless Telegraphy: To the Editor of *The Times*," *Times*, October 29, 1906. See also the articles of S. P. Thompson (October 12, 1906), Kelvin (October 16, 1906), and Swinton (October 29, 1906) under the heading of "Wireless Telegraphy" in the *Times*.

72. In the present volume, Eccles's diagram is reproduced as figure 2.1.

Chapter 3

1. "Marconi Signals across the Atlantic," *Electrical World* 38 (1901): 1023–1025.

2. "Wireless Telegraphy," *Electrical World* 38 (1901): 1011.

3. On Marconi's first transatlantic telegraphy, see Vyvyan 1933, pp. 23–33; Jacot and Collier 1935, pp. 62 ff; Dunlap 1937, pp. 87–102; Danna 1967, pp. 26–33; Clayton 1968 (ch. 7: "The First Transatlantic Wireless Messages"), pp. 133–150; W. J. Baker 1970, pp. 61–73; Jolly 1972, pp. 85–114; Süsskind 1974; Geddes 1974, pp. 14–20; Aitken 1976, pp. 261–265; D. Marconi 1982; Isted 1991b, esp. pp. 110–112. For Marconi's own accounts, see Marconi's letter in *New York Herald*, December 17, 1901; Marconi 1903a; Dunlap's interview with Marconi in Dunlap 1937, pp. 94–98, passim.

4. Among the secondary literature, only Vyvyan, Baker, and Aitken duly appraise (though very briefly and often incorrectly) Fleming's role in the Poldhu experiment. Baker and Aitken rely on Vyvyan, who had assisted Fleming at Poldhu in the winter of 1900–01 and had borrowed a manuscript titled The History of Transatlantic Wireless Telegraphy from Fleming while preparing Vyvyan 1933. Fleming's own account was published briefly in Fleming 1906, pp. 44–45, 69–70, 449–452. A yet more detailed account can be found in an unpublished manuscript, John Ambrose Fleming, The History of Transatlantic Wireless Telegraphy, volume I (manuscript narrative by Fleming covering the years 1898–1902, n.d., MS Add 122/64, Fleming Collection, University College London. Fleming's manuscript notebook of the Poldhu experiment, Notebook: Experiments at UCL and at Poldhu, UCL MS Add 122/20, Fleming Collection, is another valuable source.

5. The idea of "style" in technological action (not to be confused with "national styles" in technology) has not been discussed much by historians. For notable exceptions, see Jenkins 1984; Pitt 1988; Ferguson 1977; Hounshell 1976. For a discussion of various styles of reasoning in science, see Hacking 1992b.

6. H. R. Kempe, "Signalling Across Space by Marconi," September 15, 1896, MCA HIS 64.

7. Marconi to Preece (March 31, 1897) MCA HIS 62.

8. John Ambrose Fleming, "Memo on Marconi's System and Marconi," April 9, 1898, UCL MS Add 122/48, Fleming Collection, UCL.

9. On the life and work of Fleming, see MacGregor-Morris 1954; Hong 1994a.

10. The message was: "Glad to send you greetings conveyed by electric waves through the aether from Boulogne to South Foreland, 28 miles, and thence by postal telegraphs. —Marconi." (*Times*, March 30, 1899)

11. Some authors have suggested that given Fleming's experience in power engineering, Marconi chose Fleming as scientific advisor because he anticipated expertise on how power could be increased for transatlantic wireless telegraphy. See W. J. Baker 1970, p. 63, and p. 403. I have not found, however, any evidence to support this claim, which is incompatible with Marconi's attempt in 1898 to invite Kelvin as Scientific Advisor and with Fleming's recollection that Marconi began to consider transatlantic wireless telegraphy at the end of 1899. For Kelvin's "rejection" of the position of scientific advisor to Marconi, see S. P. Thompson 1910, 2, p. 1006.

12. John Ambrose Fleming, "Wireless Telegraphy: To the Editor of *The Times*," *Times*, April 3, 1899.

13. Oliver Lodge to John Ambrose Fleming, April 11, 1899, UCL MS Add 122/66, Fleming Collection; Fleming to Lodge, April 14, 1899, MS Add 89/36, Lodge Collection, UCL.

14. (Copy of) Fleming to Jameson Davis, May 2, 1899, MS Add 122/47, Fleming Collection, UCL.

15. (Copy of) Fleming to Jameson Davis, May 2, 1899, MS Add 122/47, Fleming Collection, UCL.

16. For Lodge's magnetic induction telegraphy, see Fleming to Davis, August 19, 1899, MCA; also Lodge 1898b. To Fleming's Volta-centenary lecture at Dover, "Lodge would not second the vote of thanks." On this episode, see Jolly 1972, p. 58. On Fleming's lecture and demonstration, refer to Fleming 1899b; Fleming 1934, p. 118.

17. John Ambrose Fleming, "Report to G. Marconi Esq. on Experiments Made on Relays, During the Last Four Months" (typewritten report submitted to the Marconi Company, 20 Mar 1900), 6 ff, MCA. See also Fleming to Marconi, January 15, 1900, MCA; Fleming to Marconi, February 9, 1900, MCA.

18. John Ambrose Fleming, "A Few Notes on No. 7777 of 1900," (typewritten report submitted to the Marconi Company, n.d.) 6 ff, MCA. See also Guglielmo Marconi, "Improvements in Apparatus for Wireless Telegraphy," British Patent Specification 7,777 (1900). For details of Marconi's syntonic patent, refer to chapter 4.

19. On his earlier conceptions, see G. Marconi, "Improvements in Transmitting Electrical Impulses and Signals, and in Apparatus Therefor," British Patent Specification 12,039 (Provisional Specification, June 2, 1896). Marconi's conception of the earth as waveguide seems to have originated from the influence of John Fletcher Moulton, who helped Marconi with the complete specification of his first patent in 1896. Moulton studied mathematical physics at Cambridge, and certainly knew about the Maxwellian theories of waveguides as described by J. J. Thomson 1893. For Moulton's possible influence on Marconi, see Silvanus Thompson to Oliver Lodge, June 30, 1897, UCL MS Add 89, Lodge Collection. For the British Maxwellian's notion of waveguide, see Buchwald 1994, pp. 333–339.

20. Even if the Marconi's wave were assumed to act as a guided wave, it would differ from a wave guided by, say, two parallel wires, since there was only one means

of wave-guidance (i.e. the earth) in Marconi's case. Because of this his long-distance transmission had to essentially depend on the power of radiation and the sensitivity of detectors.

21. For Marconi, see "Marconi's Recent Work in Wireless Telegraphy," *Electrical World* 33 (1899): 608. For Silvanus Thompson's comment, see *Electrical World* 33 (1899): 444. Thompson had argued for long-distance wireless telegraphy in S. P. Thompson 1898, p. 459. Thompson did not give any reason why the wave could reach below the earth's curvature, but George FitzGerald thought that diffraction of the long-wave at the earth's edge made it travel around the world. For this, see George FitzGerald to Oliver Heaviside, May 7, 1899, in Nahin 1988, p. 273. For the interview of Marconi's friend by the *Pall Mall Gazette*, see *Electrical World* 33 (1899): 583.

22. Fleming to Marconi, August 23, 1899, MCA.

23. Fleming, History of Transatlantic Wireless Telegraphy, volume I (hereafter cited as Fleming, History).

24. For the company's hard financial situation and the tension between Marconi and the board, see W. J. Baker 1970, pp. 62–63; Geddes 1974, p. 14. At that time, *Electrical Review* noticed that Marconi's wireless telegraphy was certainly practical, but had not returned profits to its investors; "The Commercial Possibility of Wireless Telegraphy," *Electrical Review* 46 (1900): 337–338. Marconi's syntonic demonstration to the board members is described in J. A. Fleming, "Recent Advances in Wireless Telegraphy: To the Editor of *The Times*," *Times*, October 4, 1901. Even in September 1900, Flood-Page tried to persuade Marconi to first experiment with transmission over a moderate distance, e.g., between England and Spain. Flood-Page, "Memorandum," September 19, 1900, MCA.

25. Fleming to Marconi, May 3, 1900, MCA.

26. Fleming to Flood-Page, July 2, 1900, in Fleming, History, on p. 7. Though Fleming conjectured that "for this outlay [£1,000] we shall have a plant that will enable us to settle the question of very long distance telegraphy," the total cost spent for this experiment turned out to be £50,000.

27. Fleming to Flood-Page, July 18, 1900, in Fleming, History, p. 11.

28. Each condenser consisted of 12 glass plates (16 × 16 inches) alternated with zinc plates, all of which were immersed in a wooden box filled with linseed oil. The condenser was designed to endure extremely high voltage. See Fleming, History, p. 14.

29. If the 2-inch spark corresponded to 100,000 volts, its power stored on a 0.02 microfarad condenser would be roughly 30–50 kW, within range of an alternator frequency of 30–50 Hz. The system needed as much energy as possible to be stored in the condenser. Though the energy stored in a charged condenser is proportional to its capacitance and the applied voltage ($E = CV^2/2$), the capacitance could not, however, be increased over a certain limit (say, 0.033 microfarad) because of the difficulty in tuning the discharge circuit with an antenna of much smaller capacitance; the alternative was to increase the applied voltage.

30. Fleming, History, p. 24. Fleming first conceived of the double-transformation system in July 1900. See Flood-Page to Fleming, July 25, 1900, in ibid., p. 19. For Marconi's thought on the 2-inch spark, see ibid., p. 23.

31. The use of two condensers of different capacitances was an essential feature of Fleming's double-transformation system. I thank Jed Buchwald for his help in clarifying this point. For the contemporary recognition of Fleming's double-transformation system, see Poincaré and Vreeland 1904, p. 154.

32. On Fleming's work in power engineering, see Hong 1995a,b.

33. John Ambrose Fleming (with Marconi's Wireless Telegraph Company), "Improvements in Apparatus for the Production of Electrical Oscillation," British Patent Specification 18,865, October 22, 1900. The employment of an alternator and a transformer to create high-frequency oscillations had been tried by Elihu Thomson in the 1890s. Fleming's design of the revolving-arm mechanism was similar to Thomson's design in several ways, and is likely to have been influenced by the latter. For Elihu Thomson's system for creating powerful oscillations, see E. Thomson 1899.

34. Fleming to Marconi, November 9, 1900, MCA.

35. Fleming to Marconi, November 14, 1900, MCA. See also John Ambrose Fleming (with Marconi's Wireless Telegraph Company), "Improvements in Apparatus for Signalling by Wireless Telegraphy," British Patent Specification 20,576, November 14, 1900.

36. John Ambrose Fleming (with Marconi's Wireless Telegraph Company), "Improvements in Apparatus for Wireless Telegraphy," British Patent Specification 22,106, December 5, 1900.

37. John Ambrose Fleming (with Marconi's Wireless Telegraph Company), "Improvements in Methods for Producing Electric Waves," British Patent Specification 24,825, December 5, 1900.

38. Fleming to Marconi, November 26, 1900, MCA.

39. First draft of Fleming to Flood-Page, November 23, 1900, UCL MS Add 122/47, Fleming Collection.

40. Second draft of Fleming to Flood-Page, November 23, 1900, UCL, MS Add 122/47, Fleming Collection.

41. Fleming, History, p. 27. On Fleming's request, see Flood-Page to Marconi, November 29, 1900, MCA; Flood-Page to Fleming, December 1, 1900, UCL MS Add 122/47, Fleming Collection; Fleming to Flood-Page, December 3, 1900, UCL MS Add 122/47, Fleming Collection. For Fleming's patents, see the patent specifications in note 42 and 43.

42. Marconi to Fleming, December 10, 1900, UCL MS Add 122/47, Fleming Collection.

43. Fleming to Marconi, December 13, 1900, MCA.

44. Diary of G. S. Kemp (1900–01), III, p. 153, typewritten manuscript, MCA.

45. John Ambrose Fleming, *Notebook: Experiments at UCL and at Poldhu*, January 26–29, 1901, UCL MS Add 122/20, Fleming Collection (hereafter cited as Fleming, Notebook).

46. On February 15, 1901, Fleming wrote to Marconi: "I am yet uncertain as to whether there will be much or little difficulty in obtaining the 2 inch oscillatory spark you require." (MCA)

47. John Ambrose Fleming (with Marconi's Wireless Telegraph Company), "Improvements in Apparatus in Wireless Telegraphy," British Patent Specification 3,481, February 18, 1901).

48. Fleming, History, p. 29. See also Fleming to Marconi, February 27, 1901, MCA.

49. John Ambrose Fleming, "Recommendations with Regard to the Marconi Electric Power Station in the United States," February 20, 1901, p. 8 ff; "Supplementary Recommendation with Regard to the Alternator for the USA Station," (1 Mar 1901), 2 ff, MCA. Kemp, Diary, p. 160.

50. Fleming to Marconi, February 19, 1901, MCA; Fleming to Marconi, February 27, 1901, MCA. Tesla had planned the transatlantic transmission of signals since 1899, but began to construct the famous Wardenclyffe tower in early 1901. See "Notes," *The Electrician* 43 (1899): 144; "Tesla's Wireless Telegraphy," *Electrical Review* 48 (1901): 306. On the connection between Tesla and Morgan, see Seifer 1985.

51. Fleming, *Notebook*, April 17, 1901, Wednesday. Also in Kemp, Diary (p. 166).

52. The "wattless current" in AC was a component of the primary current that lags 90° behind the primary voltage. The other component 90° ahead of the voltage was called the "watt current." Therefore, total primary current $I = (I_{watt}^2 + I_{wattless}^2)^{1/2}$, $I_{watt} = I \cos\theta$, and $I_{wattless} = I \sin\theta$, where θ is the phase difference between the current and the voltage. For Fleming's acquaintance with such technique in power engineering, see Hong 1995a.

53. Fleming, *Notebook*, April 18–19, 1901 (unpaginated).

54. Fleming, History, p. 35.

55. Fleming, *Notebook*, May 28, 1901 (unpaginated).

56. Fleming, *Notebook*, note on June 5 (unpaginated). For Fleming's work on the condensers, see Fleming to Marconi, June 3, 1901; June 13, 1901, MCA. On the communication between Poldhu and St. Catherine's, Fleming to Marconi, June 21, 1901, MCA.

57. For the Crookhaven communication, see G. S. Kemp to Marconi, June 29, 1901, MCA. Kemp had created several codes for the quality of the message received, which later came to be used more widely. MM stood for "absolutely," M for "red" or "perfect," Q for "crimson" or "good but missing," S for "yellow" or "partially readable," 1 for "green" or "strong but unreadable," 2 for "indigo" or "weak," and 3 for "violet" or "nothing." At the time, 'aerial' was the prevalent term, rather than 'antenna'.

58. For the July experiment, see Fleming, *Notebook*, "Sending to Crookhaven," July 4, 1901 (unpaginated).

59. Fleming, *Notebook*, July 5, 1901 (unpaginated).

60. "Marconi's Own Story of Transatlantic Signals" (extract from "The Weekly Marconigram" (manuscript dated June 25, 1903; MCA HIS 74, p. 2).

61. Fleming, *Notebook*, July 8, 1901, and pages under the heading of "Experiments at Poldhu July 10th 1901"; Kemp, Diary, p. 178. The various provisional specifications of Fleming's double-transformation system, filed in 1900 and early 1901, do not mention tuning at all. Only since the complete specification of his patent 20,576, filed on August 13, 1901, did Fleming begin to mention tuning between three circuits.

62. Fleming, History, p. 38.

63. Marconi also detected signals with his ordinary coherers at Crookhaven. If the new coherer were about ten times more efficient than ordinary ones, this would make 2000-mile transmission feasible (Fleming, History, p. 42). Owing to this new coherer, however, Marconi was later involved in a bitter controversy over who invented it (Phillips 1993). Phillips demonstrates that this coherer actually operated as a rectifier, unlike ordinary coherers.

64. Marconi to Entwistle, November 22, 1901, quoted in Marconi, "Marconi's Own Story of Transatlantic Signals," MS MCA.

65. "Experiments in Wireless Telegraphy," *Electrical World* 38 (1901): 990. Vyvyan (1933, p. 29) recalled that Marconi kept his true aim secret because "if he stated his purpose beforehand and failed, it would throw some discredit on his system in its more modest scope, whereas if he succeeded the success would be all the greater by reason of its total unexpectedness."

66. The letter S (that is, ···) was chosen because of the aforementioned defect in the Poldhu system whereby pressing the key long enough for a dashes created a dangerous arc across the spark gap. (See Marconi 1908, p. 114.) Just after the success in December 1901, Marconi, however, told the *New York Sun* that "the test letter is changed from week to week, and when the transatlantic message was received at Newfoundland it happened to be the turn to telegraph s" (*Electrical World* 39, 1902): 24).

67. For Marconi's own account of the reception of the signal, repeated in much of the secondary literature, see Dunlap 1937, pp. 94–98. See also D. Marconi 1982, pp. 90–94; Hancock 1974, p. 34. There are different opinions concerning the wavelength that Marconi used. In several places, Fleming estimated it as 700–1000 meters, but H. M. Dowsett, an engineer of the Marconi Company, gave 366 meters as the wavelength used in his test of June 1901. (See W. J. Baker 1970, p. 68.) The radio engineer Edwin Howard Armstrong once made an interesting comment on this issue. In his letter to H. J. Round, a former engineer at the Marconi Company, Armstrong wrote: "Apparently no one who was there gave any consistent value with respect to it, and I suppose the answer is they were guessing as we were, although probably with less knowledge of the fundamentals when they made the guess." E. H. Armstrong to H. J. Round (April 16, 1951), Armstrong Papers, Columbia University, New York City. The wavelength was not important in the actual reception of the signal. Because of the swing of the kite, which caused variation in the capacitance, Marconi abandoned the tuned system for an untuned one with a mercury coherer. A modern calculation casts doubt on Marconi's reception of signals with his untuned receiver by showing that his untuned kite (500 feet) could only have responded to frequencies greater than 5 MHz (wavelength being shorter than 60 meters). See Ratcliffe 1974a,b. For historians, however, what is

more interesting than such technical estimation is examining how the authority and authenticity of Marconi's claim was then constructed—a subject on which a further research is required.

68. For various professional and nonprofessional reactions to Marconi's claim, see Danna 1967, pp. 40–61. Elihu Thomson's strong support for Marconi contributed greatly to changing the opinions of US engineers. For this, see D. Marconi 1982, p. 103.

69. For a similar passage, see Woodbury 1944, p. 235.

70. Fleming, *Notebook*, page under heading "Decbr, 1901."

71. On Marconi's interview with the *New York Herald*, see "Marconi Signals Across the Atlantic," *Electrical World* 38 (1901): 1023–1025. On the AIEE dinner, see "The Institute Annual Dinner and Mr. Marconi," *Electrical World* 39 (1902): 124–126. For Marconi's mention of Fleming in his address at the AIEE, see Marconi 1903a, p. 99. For Marconi's speech in Britain, see Marconi 1902a. For contemporary popular reports in which Fleming's contribution was neglected, see McGrath 1902; Baker 1902.

72. "Notes," *The Electrician* 48 (1902): 761–762.

73. As is discussed in chapter 2, Thompson's main point was to prove that the inventor of wireless telegraphy was Oliver Lodge, not Marconi. On the priority dispute over the Italian Navy coherer, see Phillips 1993.

74. Marconi to Fleming, May 19, 1902, UCL MS Add 122/47, Fleming Collection.

75. (Copy of) Fleming to Marconi, May 21, 1902, UCL MS ADD 122/47, Fleming Collection.

76. For Marconi's anxiety about Fleming's alterations, see Marconi to H. Cuthbert Hall, June 29, 1902, Gioia Marconi Braga Private Collection. On the July experiment, see Fleming, *Notebook*, July 4–7, 1902. On Marconi's reception of signals on board Carlo Alberto, see Solari 1902. On Marconi's demonstration for the Russian emperor, see Aitken 1976, p. 295, n. 83.

77. Marconi to H. Cuthbert Hall, August 22, 1902, Gioia Marconi Braga Private Collection. Also in Douglas 1987, p. 36, note 14.

78. Marconi to H. Cuthbert Hall, August 22, 1902, Gioia Marconi Braga Private Collection. For Marconi's anger toward Fleming, see Marconi to H. Cuthbert Hall, October 2, 1902, Gioia Marconi Braga Private Collection.

79. Marconi to Fleming, February 2, 1903; (copy of) Fleming to Marconi, February 10, 1903, UCL MS Add 122/47, Fleming Collection.

80. Fleming to Lodge, August 29, 1937, in UCL MS Add 89/36 Lodge Collection.

81. On Fleming's education and research in Cambridge, see Hong 1994a, section 1.2.

82. For the best description of Marconi's methodology, see Aitken 1976, pp. 179–297, passim. Among Marconi's own statements, Marconi 1901 is the most important source that shows Marconi's creativity to design. See also Fleming 1937, p. 57.

83. "Although, easy to describe," Fleming once noted (1900, p. 90), "it requires great dexterity and skill to effect the required tuning [with Marconi's jigger]."

84. Ferguson (1977) characterizes engineers' intiutive method and skills in terms of "the mind's eye." On Watt, Brindley, and Brunel, see Smiles 1904 and Hughes 1966. On Crompton, see Bowers 1969. On Ferranti, see Hughes 1983, pp. 237–246.

85. Marconi to Fleming, May 19, 1902, UCL MS Add 122/47 Fleming Collection.

86. On Fleming's establishment of professional credibility, see Hong 1995a,b.

Chapter 4

1. The phenomenon of multiple resonance was first discovered by the Swiss physicists E. Sarasin and Lucien de la Rive. On the basis of their discovery, they claimed that the wave generated by the Hertzian device was a composite of heterogeneous waves (as white light is a composite of various heterogeneous waves) dispersed throughout a very broad frequency range. Hertz attributed this anomaly to the damping of a wave with a definite frequency. For a discussion of multiple resonance, see Aitken 1976, pp. 70–73.

2. Lodge to Thompson, April 14, 1897, University College London (UCL).

3. Oliver Lodge, "Improvements in Syntonized Telegraphy without Wires." British Patent 11575 (application May 10, 1897; complete specification February 5, 1898; accepted August 10, 1898).

4. Guglielmo Marconi, "Improvements in Transmitting Electrical Impulses and Signals, and in Apparatus Therefor." British Patent 12039. Date of application, June 2, 1896; Complete specification, March 2, 1897; accepted July 2, 1897. On November 10, 1896, Marconi wrote: "Ethereal vibrations effect the conductors at the receiver (which ought to be *electrically tuned* with the transmitter)." (Marconi to Preece, November 10, 1896, MCA HIS 62)

5. Lodge's paper was not published in the *Proceedings of the Physical Society*, but it was fully described in Lodge 1898a.

6. After the contents of Lodge's syntonic patent became known to the public, however, Marconi erased these two statements when filing his Complete Specification. Marconi thought that his "inductive-coupling" in receivers "not only improves the signals but also prevents to a great extent any interference due to atmospheric influence." Marconi, "Improvements in Apparatus Employed in Wireless Telegraphy," 12326 (Provisional Specification, June 1, 1898).

7. For a description of the experiment, see Marconi 1901, p. 509.

8. Diary of G. S. Kemp, volume 1 (1897 to 1898), p. 41ff (MCA).

9. Lodge (1894, p. 325) explained: "If the coatings of the jar are separated to a greater distance, so that the dielectric is more exposed, it radiates better. . . . By separating the coats of the jar as far as possible we get a typical Hertz vibrator, whose dielectric extends out into the room, and this radiates very powerfully."

10. G. Marconi, "Improvements in Apparatus for Wireless Telegraphy," 5387 (application March 21, 1900; complete specification January 21, 1901; accepted June 21, 1901).

11. G. Marconi, "Improvements in Apparatus for Wireless Telegraphy," British Patent 7,777 (application April 26, 1900; complete specification February 25, 1901; accepted April 13, 1901)

12. If one solves the series of complex differential equations that govern such a coupled circuit (as V. Bjerknes had done 5 years before Marconi's technological achievement), one finds that the system, given a certain condition, generates nearly continuous waves. See Bjerknes 1895.

13. Diary of G. S. Kemp, volume 2 (1899 to 1900), pp. 105–111 (MCA).

14. In early 1901, Marconi succeeded in transmitting 180 miles with this syntonic wireless system between St. Catherine's and the Lizard.

15. F. Braun, "Improvements relating to the Transmission of Electric Telegraph Signals without Connecting Wires." British Patent 1862 (application January 26, 1899; complete specification, October 23, 1899; accepted January 6, 1900.

16. Fleming to Marconi, February 20, 1901.

17. John Ambrose Fleming, "A Few Notes on No. 7777 of 1900," (written in February 1901), MCA.

18. Fleming to Marconi, June 9, 1911, MCA.

19. See chapter 3 above.

20. Fleming, "Recent Advances in Wireless Telegraphy: To the Editor of *The Times*." *Times*, October 4, 1900. Fleming mentioned the experiments again in an influential Cantor Lecture to the Society of Arts (Fleming 1900, pp. 90–91).

21. *Electrical World* 38 (1901), October, p. 596.

22. *The Electrician*, January 17, 1902.

23. *The Electrician*, February 28, 1902.

24. This announcement was quite unusual, since the patent for the magnetic detector was not fully granted. Marconi had filed only a provisional specification in May 1902. Never before had Marconi publicized an invention before his complete specification had been accepted.

25. *The Electrician* 50 (1902), October 24.

26. Ibid.

27. *Times*, September 22, 1901.

28. I was not able to locate a reliable biographical source on Nevil Maskelyne. The information in this paragraph is collected from various engineering journals and newspapers of the day. Jolly's biography of Marconi (Jolly 1972) mentions Maskelyne's collaboration with Hozier for Lloyd's.

29. *The Electrician*, November 7, 1902.

30. *The Electrician*, November 21, 1902.

31. "Anglo-American Cable: Wireless Competition," *Telegraph* (February 3, 1903); "Wireless Telegraphy's Future: Mr. Marconi Replies to the Croakers—Can Message be Tapped?," *St. James Gazette* (February 9, 1903); Nevil Maskelyne, "The 'Tapping' of Marconigrams," *St. James Gazette* (February 11, 1903).

32. Fleming, "Power Stations and Ship-To-Shore Wireless Telegraphy: Letter to the Editor of *The Times*." *Times*, April 14, 1903.

33. For a discussion of witnessing and credibility in early modern science, see Shapin and Schaffer 1985.

34. *Daily Telegraph*, March 25, 1903.

35. *The Electrician*, May 29,1903, p. 235.

36. *The Electrician*, June 12, 1903, p. 315.

37. See Maskelyne 1903a, p. 358. There is an interesting collection on the Maskelyne affair kept at the IEE. It is a collection of newspaper and magazine clippings published in 1903 concerning Maskelyne's jamming. *Wireless Telegraphy* (collected by Emily J. Sharman, 1903) S.C. Mss 17, IEE, London.

38. The message that Maskelyne transmitted, "rats," carried a cultural meaning. During the Boer War (1899–1902), the British Army heavily bombarded a Boer entrenchment, and asked them, by means of a heliograph, what they felt about the power of British shells. The answer the Boer army sent back, by means of heliograph, was "rats." After this, the word came to symbolize "a warning against overwhelming pride" (Austin 1903).

39. Blok's recollection differs a bit from Maskelyne's interview with the *Daily Express* soon after the affair ("Ghost's Tapping," *Daily Express*, June 13, 1903). Maskelyne said that an observer of his side noticed that, after the inception of "rats" messages, "the operator tore off the tape, rolled it up into a ball, and threw it away." Maskelyne then stated that "Fleming had the Morse instrument switched off, and the subsequent messages were received through a telephone." He admitted that he sent few lines from Shakesphere's *Henry V* and *The Merchant of Venice*, but never mentioned the derisive doggerel.

40. According to Jacot and Collier (1935), Nevil Maskelyne's son, Jasper, recalled that Maskelyne bet Fleming that he would upset the meeting. However, I have not found any evidence that supports this claim.

41. Fleming to Marconi, June 5, 1903, MCA.

42. Fleming to Marconi, June 6, 1903, MCA.

43. "Scientific Hooligans," *Daily Telegraph*, June 12, 1903. A 15-year-old boy with the surname Bruce had apparently assisted Maskelyne at the time. In 1959 he asserted categorically that genuine radio transmissions had been made from the rooftop of the Egyptian theatre (F. Shore to G. G. Hopkins, April 22, 1959, MCA).

44. John Ambrose Fleming, "Wireless Telegraphy at the Royal Institution: To the Editor of *The Times*." *Times*, June 11, 1903.

45. "Scientific Hooligans," *Daily Telegraph*, June 12, 1903. See also interview with C. Hall in "Wireless Rats," *Morning Leader*, June 12, 1903.

46. Nevil Maskelyne, "Wireless Telegraphy: To the Editor of *The Times*," *Times*, June 13, 1903.

47. "Ghost's Tapping," *Daily Express*, June 13, 1903.

48. Fleming, "Wireless Telegraphy at the Royal Institution: To the Editor of *The Times*," *Times*, June 16, 1903. Here Fleming mentioned that Maskelyne "made a gratuitous and erroneous assumption that the apparatus I was using was a syntonic apparatus" and "the experiments were in no sense whatever an exhibition of the reliability and efficacy of Marconi syntony."

49. "Scientific Hooliganism," *St. James's Gazette*, June 13, 1903; Nevil Maskelyne, "Wireless Telegraphy at the Royal Institution," *Times*, June 18, 1903.

50. Charles Bright, "To the Editor of *The Times*," *Times*, June 16, 1903.

51. *Electrical Review*, June 19, 1903.

52. "New Wireless Wonder: Questions for Mr. Marconi," *Morning Advertiser*, July 10, 1903.

53. "The Daily Wireless," *Punch* 124 (1903), July 1, p. 453.

54. The reason why Fleming initially asked for such a modest salary in 1899 was that he regarded his advisorship to Marconi as a part-time position for which he would not expend much energy or time. See (copy of) Fleming to Jameson Davis, May 2, 1899, UCL. See also chapter 3 above.

Chapter 5

1. See, e.g., Sharp 1921; White 1943; Shiers 1969.

2. On the discovery of the phantom shadow in Edison's Menlo Park laboratory, see W. J. Hammer, Memoranda, dated 3 Oct. 1884, 6 pages, MS, Hammer Collection, Smithsonian Institution, Washington DC; F Jehl, *The "Edison Effect" Tube*, Hammer Collection. See also Josephson 1959, pp. 274–275.

3. On Edison's utilization of the effect, see Johnson 1960, pp. 766–770; White 1943; Thackeray 1984. For Edison's patent for the voltage regulator, see T. A. Edison, "Electrical Indicators," U.S. Patent 307,031, October 21, 1884.

4. Sources of quotations: Johnson 1960, p. 767; "A Phenomena of the Edison Lamp,"*Engineering* 38, 1884: p. 553; Houston 1884, p. 2. However, Houston (ibid., p. 3) also wrote: "I should say here that I am not entirely convinced from the few experiments I have tried myself, as to the actual existence of such a current [in the negative connection]. The deflection of the galvanometer needle, when connection was made with the negative terminal, being quite feeble. I am assured, however, that decided deflections have been observed."

5. William Crookes argued that radiant matter, projected from the surface of the cathode, consisted of cathode rays. The radiant matter, which was neither solid, liquid, nor gas, was negatively charged and traveled with enormous velocity. On Crookes's radiant matter, see Crookes 1879; Woodruff 1966; DeKosky 1976. Before Crookes, Cromwell Varley in Britain (in 1871) and Eugen Goldstein in Germany (in 1876) suggested that the cathode ray consisted of negatively charged particles of matter. See Anderson 1964.

6. William Spottiswoode conducted a series of experiments on the discharge in rarefied gases in the late 1870s and the early 1880s. He concluded in 1880 that

"negative electricity in the tube . . . outruns the molecular streams" (Spottiswoode 1880).

7. Oliver Lodge read a paper on the creation of fog at the meeting of the British Association for the Advancement of Science in September 1884, held in Montreal, that Preece had attended before his visit to the United States.

8. See Preece 1885. On Preece's work on the Edison effect, see also Tucker 1981–82, pp. 124–125.

9. Preece's measurements and calculations were standard practice in telegraphy, especially when one wanted to check if a cable was cut at some place or not. See Preece 1885, tables on pp. 221–229.

10. In 1884, Fleming visited the US after the annual meeting of the British Association in Montreal, and met Edison in New York. Even though Fleming examined Edison's Pearl Central Station, he did not apparently hear about the Edison effect from anyone. Even Fleming's second paper, read at the Physical Society in June 1885, did not mention Preece's work read at the Royal Society a few months ago. On Fleming's work in New York, see Hong 1994a, p. 63.

11. Because of the difficulty in constructing the special lamps, physicists preformed an experiment on the Edison effect with a cathode ray tube that had a carbon cathode. See, e.g., Crookes 1891a, pp. 21–25. Arthur Schuster (1890, p. 532) remarked that the Edison effect was a phenomenon which "has received some attention of *electricians* (not physicists)." Julius Elster and Hans Geitel performed a series of experiments in the 1880s on the measurement of the potential difference between a plate anode and a hot metallic cathode, both of which were enclosed in a glass filled with a rarefied gas or in vacuum; see, e.g., Elster and Geitel 1887.

12. For Fleming's proposal of the National Standardizing Laboratory, see Fleming 1885b.

13. For Fleming's research on photometric standards and carbon filaments, see J. A. Fleming, "Experiments on Carbon Filament", 1887 in *Experiments on Carbon Filaments, Voltmeters etc.* (Edison and Swan United Electric Light Co., Ponder's End), MS Add 122/8, UCL, Fleming Collection; Fleming, "Tests on Carbon Filament" (1887–1889), MS Add 122/9, UCL Fleming Collection.

14. On this basis, Fleming (1890a, p. 120) noted that "when the lamp is actuated by an alternating current a continuous current is found flowing through a galvanometer" connected between the plate and the positive pole of the filament. However, this kind of rectification had little practical use, because it merely transformed a heavy alternating current into a feeble direct current.

15. Modern vacuum technology begins with Johann Geissler, a Bonn instrument maker who invented a mercury vacuum pump in 1858. In the early 1880s, the vacuum in the Edison lamp, prepared with the Geissler pump, was usually around 10^{-6} atmospheric pressure, which was regarded as the optimum for commercial lamps. Crookes, with Sprengel's pump, created the best vacuum of 10^{-7} atm in 1876, and Gimingham used a modified Sprengel's pump of his own design when he obtained the vacuum of 10^{-8} in 1884. For the improvement of the vacuum technology in the 1880s see Thompson 1887–88. I believe that Charles Gimingham's contribution

was essential to Fleming. Gimingham, an expert on lamp construction, was a technician in the Edison-Swan Company. He was very collegial with Fleming, and supplied several special lamps for the latter's research on the Edison effect. On Gimingham's work, see DeKosky 1983, p. 17.

16. See also Guthrie 1873. Fleming later recalled that "Guthrie here made an early excursion into a field of research, viz., thermionic emission, which has had such important developments of late years" (Fleming 1924, p. 20). On Fleming's education at the South Kensington School, see Hong 1994a, pp. 13–16.

17. On the Maxwellian theory of conductivity, see Buchwald 1985, pp. 30–32. For Fleming's experiments with a condenser and a Clark Cell, see Fleming 1890b, pp. 43–44.

18. For this story, see Hong 1994a, pp. 17–21.

19. In the Maxwellian theory, as Buchwald (1985) has nicely shown, examining the conductivity of a given substance constituted a very important, yet tacit, strategy for probing its electrical character. Maxwellian theory regarded the conduction current as, in effect, a convergence of field energy onto the circuit. Maxwell himself had been much interested in Schuster's early observation (1874) of what was termed "unilateral conductivity" (coined by Schuster), which occurs when an AC source is applied to a circuit of copper wires separated by a tiny air gap. This problem was so acute, in fact, that George Chrystal, under Maxwell's direction, tried to, and in fact did, abolish Schuster's unilateral conductivity. Chrystal convincingly showed that the "unilateral conductivity" in Schuster's circuit was an instrumental artifact attributable to a property of the particular kind of galvanometer employed (Schuster 1874; Chrystal 1876). It is highly likely that Fleming was acquainted with the term while at the Cavendish Laboratory.

20. Fleming (1896, p. 231) stated that this experiment was not successful.

21. For Fleming's recollection after 1920, see Fleming 1923a; Fleming 1923b (I thank Hasok Chang for this reference); Fleming 1934, pp. 140–142.

22. On Marconi's magnetic detector, see O'Dell 1983. On electrolytic detectors, see Philips 1980, pp. 65–84. On Lodge's system, see "The Lodge-Muirhead Military Wireless Telegraph Apparatus," *The Electrician* 51 (1903), October 16: 1036–1037.

23. On Fleming's research in 1903–04, see Hong 1994a, pp. 263–265.

24. Some of the measuring instruments, such as Duddell's thermo-galvanometer, were actually used as detectors. See Duddell 1904.

25. For Rutherford's magnetic detector, see Rutherford 1897. For Marconi's magnetic detector, see Marconi 1902b. For Fleming and Blok's work, see A. Blok, *Rutherford's expt. and a modification* (October 1902) and *E. Mag. effects of Oscillating currents* (November 1902), in Laboratory Book, vol. I, UCL MS Add 122/21. Fleming Collection, pp. 74–90.

26. See also Fleming 1903a, p. 762.

27. For Fleming's own comments on his device, see Fleming 1906a, pp. 383–385.

28. AC dynamometers, which were calibrated to read the root-mean-square value of current i, actually measured the square value i^2. Thus, if the current falls from 1

to 10⁻³ ampere (1 milliampere), the scale indicated by the needle of the instrument falls from a division of scale corresponding to 1 ampere to its 10⁻⁶ value. See Duddell 1904, p. 91.

29. See chapters 3 and 4 above.

30. For Fleming's engineering style in the 1880s and the 1890s, see Fleming 1934, p. 112. For an analysis of the invention of the cymometer, see Hong 1994b.

31. For a useful description of the rectifiers, see Rosling 1906. The vacuum tube rectifier refers to the gaseous vacuum tube rectifiers designed by an American physicist, P. G. Nutting, for high-voltage AC in 1904. See "Gaseous Rectifiers," *The Electrician*, 53, 1904: p. 639.

32. On Lodge's use of mercury lamp rectifiers in wireless telegraphy, see Lodge 1905, p. 84.

33. See "Schlömilch Wave Detector," *The Electrician* 53 (1904), p. 755.

34. On Fleming's trial with the Nodon valve, see Fleming 1905, p. 476.

35. The quotation is from Fleming 1906d, p. 263. It was the first published identification of the date, but after the late 1910s he recalled that he had invented it one day in October (see e.g. Fleming 1934, p. 141). For the dispute between Fleming and de Forest over electrolytic detectors, see Fleming 1996d and de Forest 1996b.

36. The phrase "happy thought" is from Fleming 1934, p. 141. This first circuit is described in Fleming 1906c.

37. Among historians of technology, Bernard Carlson and Gorman (1990) emphasize the need to examine three aspects of a cognitive process in invention: mental models, heuristics, and mechanical representations. My focus is somewhat different from theirs since I want to explain the infiltration of various resources outside a local laboratory in the process of the invention or of the transformation of effect to artifact.

38. Fleming put forth his own version of an electron theory in 1902 (Fleming 1902).

39. See Richardson 1901, 1903.

40. Fleming's later stories give the impression that electron theory, by revealing the true mechanism of the Edison effect, was essential to the invention of the valve; see e.g. Fleming 1924, pp. 20–47.

41. See e.g. Galison 1985.

42. For Fleming's research on photometry in the Pender Laboratory, see Fleming 1902–03.

43. Fleming, "Improvements in Instruments for Detecting and Measuring Alternating Electric Currents," provisional specification, British patent 24,850, November 16, 1904.

44. The reason for this was not known with certainty at the time. Fleming (1905, p. 487) attributed it to the imbalance between the production of electrons or negative ions in the carbon electrode and their expulsion to another electrode.

45. Fleming to Marconi, n.d. (circa November 30, 1904), MCA. G. Shiers, who has suggested that Fleming invented his valve because of the urgent need for new

detectors, presented this letter of Fleming's to Marconi as a piece of supporting evidence. However, even in this letter, Fleming's emphasis was on the invention of a meter, not a detector. Fleming stated, "I have found a method of rectifying electrical oscillation, that is, making the flow of electricity all in the same direction, so that I can detect then with an ordinary mirror galvanometer. I have been receiving signals on an aerial with nothing but a mirror galvanometer and my device, but at present only on a laboratory scale. This opens up a wide field for work, as I can now measure exactly the effect of the transmitter. I have not mentioned this to anyone as yet as it may become very useful." The phrase that "I can now measure exactly the effect of the transmitter" is particularly noteworthy. See Shiers 1969, p. 111. Aitken (1985, pp. 211–212) also quotes this letter from Shiers, but Aitken senses that Fleming invented the valve as a meter, not as a detector.

46. Marconi to Fleming, February 14, 1905, UCL MS Add 122/1, Fleming Collection. Fleming to Marconi, February 15, 1905, MCA. Marconi 1905.

47. Fleming to Marconi, March 24, 1905, MCA.

48. *Memorandum of Agreement between the Marconi Company and Fleming*, May 26, 1905, MCA. After the new contract, Fleming wrote a complete specification of his invention of the valve for wireless telegraphy. See Fleming, Improvements in *Instruments for Detecting and Measuring Alternating Electric Currents*, complete specification, British Patent 24,850, August 15, 1904 (accepted September 21, 1905).

49. Fleming to Marconi, October 9, 1906, MCA.

50. (Copy of) Fleming to A. R. M. Simkins (Work Manager of Marconi Company), January 17, 1907, UCL MS Add 122/48, Fleming Collection; Marconi to Fleming, April 14, 1907, UCL MS Add 122/48, Fleming Collection; (copy of) Fleming to Marconi, April 15, 1907, UCL MS Add 122/48, Fleming Collection.

51. (Copy of) Fleming to Marconi, 8 Sep. 1908, UCL MS Add 122/48, Fleming Collection; Marconi to Fleming, January 8, 1909, UCL MS Add 122/48, Fleming Collection.

52. Fleming to Marconi, 9 Oct. 1906, MCA.

53. For an abstract of de Forest 1906a, see *The Electrician* 56, 1906: pp. 216–218.

54. (Copy of) Fleming to Marconi, January 14, 1907, UCL MS Add 122/48, Fleming Collection.

55. See also Price 1984; Laudan 1984; Rosenberg 1992.

56. Note that Fleming told his story of the invention of the valve in the chapter "My Scientific Researches and Wireless Inventions" in Fleming 1934. See also Shiers 1969.

57. See White 1943; Josephson 1959; Howe, 1955, p. 15.

58. I have borrowed the term "unarticulated" from Buchwald (1993).

Chapter 6

1. As in any other area of human history, there were tragic elements: Edwin Howard Armstrong, who first devised the feedback circuit for the oscillating audion,

eventually lost a 20-year legal battle with Lee de Forest. One day in 1954, sick and tired of prolonged patent litigations, Armstrong walked out the window of his 13th-floor New York apartment.

2. See, e.g., Boucheron 1920 (titled "The Vacuum Tube—An Electrical Acrobat"), in which the audion's six different functions—as a detector, detector-amplifier, detector and two-step amplifier, generator of oscillations, telephone transmitter, and telephone rectifier—are detailed. The audion's inventor, de Forest, was called by the *Scientific American* "a Merlin of Today" (Claudy 1920).

3. The audion was also called the triode. The term "triode revolution" appears in Eccles 1930.

4. For a critical analysis of the discourse that technology has its own life, see Hong 1998.

5. Capacitance largely determines the amount of energy that can be stored and dissipated from the circuit. Inductance is an electrical variable that had been known to electrical engineers as an electric momentum or inertia. Because of inductance, electricity behaves as if it had an inertia which makes it resist sudden change in its motion.

6. Here R denotes the total resistance, which consists of R_{ohmic} and $R_{radiation}$.

7. If L is infinite, then the damping is zero. In this case, we have an infinitely long wave. Marconi once adopted this method to increase the inductance of his antenna, but he soon found it to be impractical. See chapter 4 above.

8. On Marconi's disk discharger, see Aitken 1976, pp. 277–278.

9. Lodge, *The Electrician* 28 (January 29, 1892), p. 330.

10. Ayrton's paper was titled "The Variation of the Potential Difference of the Arc with Currents, Size of Carbons and Distance Apart." See *The Electrician*, September 1893. On Ayrton, see Gooday 1991.

11. Hertha Ayrton (1854–1923) is now well known as the first woman proposed to be a fellow of the Royal Society in 1902. She went to Cambridge University after taking the Cambridge University Examination for Women. She married William Ayrton in 1885, 2 years after Ayrton's first wife, Matilda, had died. Hertha was the first woman to read a paper at the Institution of Electrical Engineers ("The Hissing of the Electric Arc" in 1899) and her paper, "The Mechanism of the Electric Arc," was read before the Royal Society in 1901 by John Perry, William Ayrton's close friend. In 1902, she was proposed as a candidate for the fellowship but was rejected because she was a married woman. For her life and work, see Tattersall and McMurran 1995; Mason 1991.

12. Thompson's formula for the relationship between voltage and current in electric arc is $V = a + (bL/I)$, where V, I, and L are respectively voltage, current, and the length of the arc and where a and b are constants. Ayrton's formula was $V = a + bL + \{(c + dL)/I\}$, where a, b, c, and d are constants.

13. Thompson's reasoning is that if $R = (a/C) + R_0$ then $(dR/dC) \times C$ is negative. See also *The Electrician*'s comment in "Negative Resistance," *The Electrician*, June 26, 1896) p. 273.

14. *The Electrician*, July 31, 1896, p. 452.

15. See Poulsen 1905.

16. On Poulsen, see Aitken 1985, pp. 117–122. See also Poulsen 1905; 1906. For the level of complication that Marconi's transmitter reached just before the audion revolution, see "The Marconi System of Wireless Telegraphy," *The Electrician* (April 26, 1912), 95–98.

17. Ernst Ruhmer and Adolf Pieper, "A Process for Generating Permanently Undamped Electrical Oscillations," German Patent 173,396 (1904); F. K. Vreeland, "Apparatus for the Production and Utilization of Undamped or Sustained Electrical Oscillations," US Patent 829,934 (1906); S. Eisenstein, "Apparatus for the Production of Undamped Electrical Vibrations," US Patent 921,526 (1906).

18. "The Poulsen System of Wireless Telegraphy," *The Electrician* 58 (1906), November 30: 237.

19. *The Electrician* 58 (1906), November 16: 157.

20. "The Arc in Wireless Telegraphy," *The Electrician* 58 (1906), December 21: 374–376. The reasons for my judgement that this article was written by Fleming or at least by someone who was assisted by Fleming are as follows: first, the author's account for the historical development of the continuous wave system was very similar to that which appeared in Fleming 1906a; second, the author's discussion of Elihu Thomson's 1893 patent, on the basis of which Thomson maintained that he was the first inventor of the continuous wave system, was the same as Fleming's; third, the author referred to Fleming's lesser-known research on the unilateral conductivity of the AC arc; finally, the author's conclusion on the nature of the Poulsen arc anticipated Fleming's conclusion in June 1907.

21. Fleming to Marconi, January 1, 1907, MCA.

22. On Marconi's rotating discharger, see Baker 1970, pp. 117–119 and Aitken 1976, pp. 277–279. See also Marconi to Fleming, April 14, 1907, MCA.

23. Fleming to Marconi, February 5, 1907, MCA. Fleming's experiments at UCL were performed by his assistant, W. L. Upton.

24. Fleming's experiments were recorded in the "Experiments with Poulsen Apparatus (mainly designed for a R.I. Lecture on May 24th 1907)," in *Laboratory Book* volume III, UCL MS Add 122/22.

25. Fleming 1907b, pp. 692–695.

26. On de Forest's audion, see Chipman 1965; Aitken 1985, pp. 162–224, 233–241. On his life, see Lubens 1942 and Hijiya 1993. De Forest 1950, though somewhat exaggerated, is also useful.

27. Lee de Forest, "Oscillation-Responsive Device," US Patent 979,275 (application filed February 2, 1905; patent granted December 20, 1910).

28. Ibid.

29. Lee de Forest, "Oscillation-Responsive Device," US Patent 867,876 (application filed February 2, 1906; patent granted October 8, 1907).

30. Lee de Forest, "Oscillation-Responsive Device," US Patent 824,637 (application filed January 18, 1906; patent granted June 26, 1906).

31. De Forest, "Oscillation-Responsive Device," US Patent 824,637 (application filed January 18, 1906; patent granted June 26, 1906). One of the devices of which he drew diagrams was Figure 6.5c. See also his two other patents under the same title, 836,070 and 836,071, applications for which were filed on the same date.

32. Lee de Forest, "Static Valve for Wireless-Telegraph Systems," US Patent 823,402 (application filed December 9, 1905; patent granted June 12, 1906). Gerald Tyne, who knew de Forest, de Forest's assistant C. D. Babcock, and H. W. McCandless who owned H. W. McCandless & Co—a lamp making company that later constructed the audion for de Forest—wrote that de Forest's assistant C. D. Babcock ordered McCandless to construct "Fleming's valve" in late 1905. Tyne, however, did not quote any source for this. See Tyne 1977, p. 53. De Forest's patent on the static valve is also analyzed in Tyne 1977, p. 57.

33. De Forest did not specify the poles of batteries here, but plate I was apparently connected to the positive side of the local battery H.

34. I believe that the idea of the audion as a relay was introduced by de Forest in order to highlight and justify the difference between his and Fleming's devices—that is, relay v. valve. In de Forest 1920, however, he claimed that he had been interested in relays since 1900, and had conceived of the action of flames in the same term, which, I think, is misleading.

35. De Forest's early audion was a "soft" (or low-vacuum) audion, in which gas ionization took place. Langmuir discovered later that for the audion to be used effectively as a feedback amplifier and oscillator, it should be "hard" (or high-vacuum). Historians were divided. Aitken (1985, p. 219) was favorable to de Forest's understanding of the audion's action; Chapman (1965, p. 98), among others, argued that de Forest misunderstood its action. We may consider his "gas ionization" theory to be right, in the sense that ionization in fact occurred in the low-vacuum audion. De Forest's claim on amplification may also be correct, since his early audion amplified signals (a little, not as much as he thought). However, signals were heard on a telephone not as a result of the audion's relay or trigger action but as a result of rectification of signals. He was wrong about this point.

36. The name "audion" was coined at the suggestion of de Forest's assistant, C. D. Babcock. It derived from the Latin verb *audire* that means to hear and *ion* from the Greek *ienai*, meaning to go (de Forest 1950, p. 214; McNicol 1946, p. 164). De Forest first used the term audion in his patent on "Wireless Telegraphy," US Patent 841,386 (application filed August 27, 1906; patent granted January 15, 1907), on p. 3, to designate the "sensitive element" in the gaseous conducting medium. It soon began to denote the device itself. Pupin ("Discussion," in de Forest 1906b, p. 766) criticized the name as a "mongrel," because "it is a Latin word with a Greek ending."

37. De Forest, US Patent 841,386. On p. 2 of this patent specification, de Forest remarked: "It is preferable that the positive pole of the battery B [the local battery] should be connected to the electrode F' [the plate], and better results are obtained if the negative pole of the battery B be connected to the end of the filament F to which the positive pole of the battery B' is connected." Various kinds of control electrodes were used by physicists to control the flow of electrons and ions in cathode-ray tubes (Aitken 1985, p. 217).

38. Lee de Forest, "Device for Amplifying Feeble Electrical Currents," US Patent 841,387 (application filed October 25, 1906; patent granted January 15, 1907); McNicol 1946, p. 165.

39. See also Tyne 1977, p. 61.

40. Lee de Forest, "Space Telegraphy," US Patent 879,532 (application application filed January 29, 1907; patent granted February 18, 1908); Chapman, p. 99.

41. For the history of feedback, see Bode 1964.

42. Lowenstein is famous for his invention of the "grid-bias" amplifier.

43. Lowenstein called the audion an "ion-controller."

44. Miessner (1964, p. 23) later testified on Lowenstein's work in the "four-party" interference proceedings, but he claims "I was shut off as none of the contestants in the court was interested, and my attempt to credit Lowenstein with that important historic first in vacuum-tube oscillators came to nought."

45. De Forest's diary, February 12, 1912; entry in his notebook, April 22, 1912, Papers of Lee de Forest, Library of Congress, Washington DC. I thank Hugh Slotten for these two items.

46. For the best analysis of de Forest's research in 1912, see Aitken 1985, pp. 233–238.

47. AT&T's goal was to use the audion amplifier as a telephone repeater. By the end of 1913, de Forest had sold 10 audion amplifiers to the Navy at $500 each. C. V. Logwoon to Earl Hanson, December 20, 1913, Armstrong Papers, Columbia University.

48. De Forest (1914) later asserted that all the new properties of the audion were due to the "one vitally essential element, . . . that second cold electrode, the 'grid electrode' of my 1907 patent application, No. 879,532."

49. For Stone's testimony, see Aitken 1985, pp. 240–241. Aitken, who is critical of Stone's testimony, states that "the implication is that Stone, consciously or unconsciously, read more into his conversation with de Forest than had originally been there" (p. 241). However, Aitken suggests that in 1912 de Forest invented more than a amplifier: "What de Forest had discovered in Palo Alto was something just as important as his original discovery of the audion or his discovery that the audion could amplify; he had discovered that, in an appropriate circuit, an audion could oscillate—that is, it could serve as a generator of continuous waves, and not merely as a detector" (pp. 238–239).

50. His early letters to Mr. Underhill (an amateur radio engineer and Armstrong's friend) in 1910 show that Armstrong was familiar with spark gaps, ordinary detectors, crystals, the quenched-spark system, the Fessenden system, as well as various theoretical treatises on wireless telegraphy. See E. H. Armstrong to Mr. Underhill, January 26 and September 19, 1910, in Armstrong Papers, Columbia University.

51. An early (undated) note of Armstrong in Armstrong Papers, Columbia University.

52. The story of Armstrong's early work is based on Lessing 1956 and on Armstrong's own testimony before the court. See *Radio Corporation of America,*

*American Telephone and Telegraph Company and De Forest Radio Company,
Petitioners vs. Radio Engineering Laboratories, INC.* Supreme Court of the United
States, 1934, pp. 830–886.

53. Lessing 1956, pp. 65–68; *Radio Corporation of America, American Telephone
and Telegraph Company and De Forest Radio Company, Petitioners vs. Radio
Engineering Laboratories, INC.* Supreme Court of the United States, 1934, pp.
830–886, passim; Armstrong 1914.

54. See E. H. Armstrong to William H. Davis, November 9, 1913), and J.
Bottomely to G. Marconi, February 3, 1914, both in Armstrong papers at Columbia
University.

55. The story of Armstrong versus de Forest here is based on Gaffey 1960;
McCormack 1934; Reich 1977. See also the patent examiner's rejection of sixteen
claims in de Forest's ultra-audion patent, National Museum of American History,
Smithsonian Institution. I thank Elliot Sivowitch for bringing the latter to my
attention.

56. In the US, interference proceedings are arranged by the Patent Office, and a
patent examiner (or examiners) decides priority. Patent litigation is legal lawsuit,
and a judge decides priority.

Epilogue

1. My current research focuses on Edwin Howard Armstrong and the revolutions
in twentieth-century radio engineering.

Appendix

1. In Fleming's terminology, the co-electron was the positive electron that was connected to every negative electron by means of the ether strain.

2. See also the note on p. 751 of Fleming 1903a.

3. It is noteworthy that Fleming shared the Maxwellians' commitment to the wave
guide in the 1880s (Buchwald 1994). The idea of the earth as a waveguide and the
generation of half-waves from Marconi's transmitter had been suggested before
Fleming by A. Blondel and R. A. Fessenden (Collins 1905, p. 33).

4. On such experiments, see Marconi 1902 and Jackson 1902. Some scientists
attributed these effects to the absorption of the wave energy by ions or electrons; see
e.g. J. J. Thomson 1902; Taylor 1903. See also Fleming, 1903a, p. 710.

Bibliography

Aitken, Hugh G. J. 1976. *Syntony and Spark: The Origin of Radio.* Wiley (second edition: Princeton University Press, 1985).

Aitken, Hugh G. J. 1978. "Science, Technology and Economics: The Invention of Radio as a Case Study." In *The Dynamics of Science and Technology,* ed. W. Krohn et al. Reidel.

Aitken, Hugh G. J. 1985. *The Continuous Wave: Technology and American Radio, 1900–1932.* Princeton University Press.

Aldridge, Susan. 1995. "A Prizefight on the Wireless." *New Scientist* 146 (May 20): 46–47.

Allen, Henry W. 1903. "Wireless Telegraphy at the Royal Institution: To the Editor." *The Electrician* 51 (July 17): 549.

Anderson, D. L. 1964. *The Discovery of the Electron.* Van Nostrand.

Anderson, Leland. 1980. *Priority in the Invention of Radio: Tesla vs. Marconi.* Antique Wireless Association.

Appleyard, Rollo. 1897. "Wireless Telegraphy: To the Editor." *The Electrician* 39 (October 22): 869.

Armstrong, E. H. 1914. "Operating Features of the Audion." *Electrical World* 64: 1149–1151.

Ayrton, Hertha. 1895. "The Electric Arc." *The Electrician* 34 (January–September), passim.

Ayrton, William. 1896a. "The True Resistance of the Electric Arc." *The Electrician* 37 (May 15): 93.

Ayrton, William. 1896b. "The Resistance of the Arc." *The Electrician* 37 (July 3): 321–322.

Ayrton, William. 1897. "Sixty Years of Submarine Telegraphy." *The Electrician* 42 (February 19): 546–548.

Austin, L. F. 1903. "Our Note Book." *Illustrated London News,* June 20.

Baker, E. C. 1976. *Sir William Preece, F.R.S.: Victorian Engineer Extraordinary.* Hutchinson.

Baker, R. S. 1902. "Marconi's Achievement: Telegraphing across the Ocean without Wires." *McClure's Magazine* 18: 291–299.

Baker, W. J. 1970. *A History of the Marconi Company.* Methuen.

Barrett, Ralph, and Margaret L. Godley. 1995. "Popov and Marconi." *Nature* 376 (August 10): 458.

Basalla, George. 1988. *The Evolution of Technology.* Cambridge University Press.

Bettòlo, G. B. Marini. 1986. "Guglielmo Marconi: Personal Memories and Documents." *Rivista di Storia della Scienza* 3: 447–458.

Bjerknes, V. 1895. "Ueber Electrische Resonanz." *Wiedemann's Annalen* 55: 121–169.

Black, George. 1903. "Wireless Telegraphy at the Royal Institution: To the Editor." *The Electrician* 51 (July 10): 503.

Blake, G. G. 1928. *History of Radio Telegraphy and Telephone.* Chapman.

Blok, Arthur. 1954. "Some Personal Recollections of Sir Ambrose Fleming." In *The Inventor of the Valve: A Biography of Sir Ambrose Fleming*, ed. J. MacGregor-Morris. Television Society.

Bode, H. W. 1964. "Feedback: The History of an Idea." In *Selected Papers on Mathematical Trends in Control Theory*, ed. R. Bellman and R. Kalaba. Doverz.

Bose, J. C. 1895a. "On the Determination of the Indices of Refraction of Various Substances for the Electric Ray." *Proceedings of the Royal Society* 59: 160–167.

Bose, J. C. 1895b. "On a New Electro-Polariscope." *The Electrician* 36 (December 27): 291–292.

Boucheron, Pierre H. 1920. "The Vacuum Tube—An Electrical Acrobat." *Scientific American*, January 17: 54–65.

Bowers, Brian. 1969. *R. E. B. Crompton.* Institute of Electrical Engineers.

Branly, E. 1891. "Variations of Conductivity under Electrical Influence." *The Electrician* 27: 221–222, 448–449.

Brittain, James E. 1970. "The Introduction of the Loading Coil: George A. Campbell and Michael I. Pupin." *Technology and Culture* 11: 36–57.

Buchanan, R. A. 1985. "The Rise of Scientific Engineering in Britain." *British Journal for the History of Science* 18: 218–233.

Buchwald, Jed Z. 1985. *From Maxwell to Microphysics.* University of Chicago Press.

Buchwald, Jed Z. 1993. "Design for Experimenting." In *World Changes: Thomas Kuhn and the Nature of Science*, ed. P. Horwich. MIT Press.

Buchwald, Jed Z. 1994. *The Creation of Scientific Effects: Heinrich Hertz and Electric Waves.* University of Chicago Press.

Buchwald, Jed Z. Forthcoming. *The Creation of Scientific Effects II.*

Burns, R. W. 1994. "Lodge and the Birth of Radio Communication," *IEE Review*, May: 131–133.

Buscemi, V. 1905. "Trasparenza dei Liquidi per le onde Hertziane." *Nuovo Cimento* 9: 105–112.

Carlson, W. Bernard, and Michael E. Gorman. 1990. "Understanding Invention as a Cognitive Process: The Case of Thomas Edison and Early Motion Pictures, 1888–91." *Social Studies of Science* 20: 387–430.

Chapin, Seymour L. 1971. "Patent Interferences and the History of Technology: A High-flying Example." *Technology and Culture* 12: 414–446.

Cheney, M. 1981. *Tesla: Man Out of Time.* Laurel.

Chipman, Robert A. 1965. "De Forest and the Triode Detector." *Scientific American* 212(3): 93–100.

Chrystal, G. 1876. "On Bi- and Unilateral Galvanometer Deflection." *Philosophical Magazine* 2 (December): 401–414.

Claudy, C. H. 1920. "A Merlin of Today: What the Audion of De Forest Has Done and What it May Do." *Scientific American,* May 15: 640, 542–554.

Clayton, Howard. 1968. *Atlantic Bridgehead: The Story of Transatlantic Communication.* Garnstone.

Collins, A. F. 1905. *Wireless Telegraphy.* McGraw.

Crookes, William. 1879. "On Radiant Matter." *Nature* 20: 419–423, 436–440.

Crookes, William. 1891a. "Electricity in Transitu: From Plenum to Vacuum." *Journal of the Institution of Electrical Engineers* 20: 4–49.

[Crookes, William.] 1891b. "Electricity in Relation to Science." *Nature* 45 (November 19): 63–64.

Crookes, William. 1892. "Some Possibilities of Electricity." *Fortnightly Review* 51 (February): 173–181.

D'Albe, E. E. Fournier. 1923. *The Life of Sir William Crookes.* Unwin.

Dam, H. J. W. 1897. "Telegraphy Without Wires: A Possibility of Electrical Science, II: The New Telegraphy—Interview with Signor Marconi." *McClure's Magazine* 8 (March): 383–392.

Danna, R. 1967. The Trans-Atlantic Radio Telegraphic Experiments of Guglielmo Marconi, 1901–1907. Ph.D. thesis, University of Missouri.

Dasgupta, Subrata. 1995–96. "Forgotten History: Sir Jagadis Bose and the Origins of Radio." *Transactions of the Newcomen Society* 67: 207–219.

Daumas, Maurice. 1979. "Introduction." In *A History of Technology & Invention,* volume 3. Crown.

De Forest, Lee. 1906a. "The Audion: A New Receiver for Wireless Telegraphy." *Transactions of the American Institute of Electrical Engineers* 25: 735–779.

De Forest, Lee. 1906b. "Oscillation Valve or Audion." *The Electrician* 58 (December 29): 425.

De Forest, Lee. 1914. "The Audion as a Generator of High-Frequency Currents." *The Electrician* 73 (August 28): 842–843.

De Forest, Lee. 1920. "The Audion—Its Action and Some Recent Applications." *Journal of the Franklin Institute* 190: 1–38.

De Forest, Lee. 1950. *Father of Radio: The Autobiography of Lee De Forest*. Wilcox & Follett.

DeKosky, Robert K. 1976. "William Crookes and the Fourth State of Matter." *Isis* 67: 36–60.

DeKosky, Robert K. 1983. "William Crookes and the Quest for Absolute Vacuum in the 1870s." *Annals of Science* 40: 1–18

Dieckmann, M. 1904. "Ueber den Schloemilch-Wellendetektor." *Physikalisch Zeitschrift* 5 (August 15): 529–531.

Douglas, Susan J. 1987. *Inventing American Broadcasting, 1899–1922*. Johns Hopkins University Press.

Duddell, William. 1900. "On Rapid Variations in the Current through the Direct-Current Arc." *Journal of the Institution of Electrical Engineers* 30 (1900): 232–283.

Duddell, William. 1904. "Some Instruments for the Measurement of Large and Small Alternating Currents." *Philosophical Magazine* 8 (July): 91–114.

Dunlap, Orrin E. 1937. *Marconi: The Man and His Wireless*. Macmillan.

Eccles, W. H. 1930. "Physics in Relation to Wireless." *Nature* 125: 894–897.

Eccles, W. H. 1933. *Wireless*. Thornton Butterworth.

Elster, Julius, and Hans Geitel. 1887. "Ueber die Electrisirung der Gase durch glühende Körper." *Wiedemann's Annalen* 31: 109–126.

Espenschid, Lloyd. 1959. "Discussion of 'A History of Some Foundations of Modern Radio-Electronic Technology.'" *Proceedings of the IRE* 47: 1253–1258.

Eve, A. S. 1939. *Rutherford*. Cambridge University Press.

Fahie, J. J. 1899. *A History of Wireless Telegraphy, 1838–1899*. Dodd, Mead.

Ferguson, E. S. 1977. "The Mind's Eye: Nonverbal Thought in Technology." *Science* 197: 827–836.

FitzGerald, George F. 1883. "On the Quantity of Energy Transferred to the Ether by a Variable Current." *Transactions of the Royal Dublin Society* (1883). Reprinted in *The Scientific Writings of the Late George Francis FitzGerald*, ed. J. Larmor (Longman, 1902).

FitzGerald, George F. 1890. "Electro-Magnetic Radiation." *Nature* 42: 172–175.

FitzGerald, George F. 1892. "On the Driving of Electro-magnetic Vibrations by Electro-magnetic and Electrostatic Engines." *The Electrician* 28 (January 29): 329–330.

FitzGerald, George F. 1896. "Science and Industry." Reprinted in *The Scientific Writings of the Late George Francis FitzGerald*, ed. J. Larmor (Longman, 1902).

FitzGerald, George F. 1896b. "Negative Resistance." *The Electrician* 37 (July 17): 386.

Fleming, John Ambrose. 1883. "On the Phenomena of Molecular Radiation in Incandescence Lamps." *Proceedings of the Physical Society* 5 (May): 283–285.

Fleming, John Ambrose. 1885a. "On Molecular Shadow in Incandescence Lamps." *Proceedings of the Physical Society* 7: 178–181.

Fleming, John Ambrose. 1885b. "On the Necessity for a National Standardizing Laboratory for Electrical Instruments." *Journal of the Society of Electrical Engineers and Electricians* 14: 488–501.

Fleming, John Ambrose. 1890a. "On the Electric Discharge between Electrodes at different Temperatures in Air and in High Vacua." *Proceedings of the Royal Society* 47: 118–126.

Fleming, John Ambrose. 1890b. "Problems in the Physics of an Electric Lamp." *Proceedings of the Royal Institution* 13: 34–49.

Fleming, John Ambrose. 1896. "A Further Examination of the Edison Effect in Glow Lamps." *Proceedings of the Physical Society* 14: 187–242.

Fleming, John Ambrose. 1899a. Discussion of Marconi's "Wireless Telegraphy." *Journal of the Institution of Electrical Engineers* 28 (March 2): 293.

Fleming, John Ambrose. 1899b. "The Centenary of the Electric Current, 1799–1899." *The Electrician* 43: 764–768.

Fleming, John Ambrose. 1900. "Electrical Oscillations and Electrical Waves." *Journal of the Society of Arts* 49: 69–131.

Fleming, John Ambrose. 1902. "The Electronic Theory of Electricity." *Proceedings of the Royal Institution* 17: 163–181.

Fleming, John Ambrose. 1902–03. "The Photometry of Electric Lamps." *Journal of the Institute of Electrical Engineers* 32: 119–171.

Fleming, John Ambrose. 1903a. "Hertzian Wave Telegraphy." *Journal of the Society of Arts* 51: 709–784.

Fleming, John Ambrose. 1903b. "A Note on a Form of Magnetic Detector for Hertzian Waves adapted for Quantitative Work." *Proceedings of the Royal Society* 71: 398–401.

Fleming, John Ambrose. 1904a. "On a Hot-Wire Ammeter for the Measurement of very small Alternating Currents." *Proceedings of the Physical Society* 19 (March 25): 173–184.

Fleming, John Ambrose. 1904b. "On Large Bulb Incandescent Electric Lamps as Secondary Standards of Light." *Report of the British Association for the Advancement of Science* (1904): 682–683

Fleming, John Ambrose. 1905. "On the Conversion of Electric Oscillations into Continuous Currents by Means of Vacuum Valve." *Proceedings of the Royal Society* 74: 476–487.

Fleming, John Ambrose. 1906a. *The Principles of Electric Wave Telegraphy.* Longman.

Fleming, John Ambrose. 1906b. "On the Electric Conductivity of a Vacuum." *Scientific American* suppl. 1568: 25129–25131.

Fleming, John Ambrose. 1906c. "The Construction and Use of Oscillation Valves for Rectifying High-Frequency Electric Currents." *Proceedings of the Physical Society* 20: 177–185.

Fleming, John Ambrose. 1906d. "Oscillation Valve, or Audion." *The Electrician* 58 (November 26): 263.

Fleming, John Ambrose. 1907a. "Oscillation Valve, or Audion." *The Electrician* 58 (January 4): 464.

Fleming, John Ambrose. 1907b. "Recent Contributions to Electric Wave Telegraphy." *Proceedings of the Royal Institution* 18: 677–710.

Fleming, John Ambrose. 1907c. "Some Observations on the Poulsen Arc as a Means of Obtaining Continuous Electrical Oscillations." Proceedings of the Physical Society 21: 23–26.

Fleming, John Ambrose. 1911. "Wireless Telegraphy." *Encyclopedia Britannica,* 11th edition, volume 26, pp. 529–542.

Fleming, John Ambrose. 1920. "The Thermionic Valve in Wireless Telegraphy and Telephony." *Proceedings of the Royal Institution* 23: 161–189.

[Fleming, John Ambrose.] 1923a. "Dr. Fleming on the Thermionic Valve: A Talk by the Inventor on Its Development and Its Application to Wireless." *Wireless World* 24 (February): 693–694.

[Fleming, John Ambrose.] 1923b. "How I Put Electrons to Work in the Radio Bottle." *Popular Radio* 3 (March): 175–182.

Fleming, John Ambrose. 1924a. *The Thermionic Valve and Its Developments in Radio- Telegraphy and Telephony.* Iliffe & Sons.

Fleming, John Ambrose. 1924b. The Physical Society of London 1874–1924: Proceedings of the Jubilee Meetings. Special Number.

Fleming, John Ambrose. 1934. *Memories of a Scientific Life.* Marshall, Morgan & Scott.

Fleming, John Ambrose. 1937. "Guglielmo Marconi and the Development of Radio-Communication." *Journal of the Society of Arts* 86: 42–63.

Gaffey, James R. 1960. "Certain Aspects of the Armstrong Regeneration, Superregeneration, and Superheterodyne Controversies." *Patent, Trade-mark and Copyright Journal of Research and Education* 4: 173–185.

Galison, Peter. 1985. "Bubble Chambers and the Experimental Workplace." In *Observation, Experiment, and Hypothesis in Modern Physical Science,* ed. P. Achinstein and O. Hannaway. MIT Press.

Galison, Peter. 1988. "History, Philosophy, and the Central Metaphor." *Science in Context* 2: 197–212.

Galison, Peter, and Alexi Assmus. 1989. "Artificial Clouds, Real Particles." In *The Uses of Experiment: Studies in the Natural Sciences,* ed. D. Gooding et al. Cambridge University Press.

Gannett, Elwood K. 1998. "The IEEE Medal of Honor." *Proceedings of the IEEE* 86: 1295–1297.

Garratt, G. R. M. 1974. "Marconi: The Lavernock Trials, May 1897." *Electronics & Power,* May 2: 323–326.

Geddes, Keith. 1974. *Guglielmo Marconi, 1874–1937.* H.M.S.O.

Gooday, Graeme. 1991. "Teaching Telegraphy and Electrotechnics in the Physics Laboratory: William Ayrton and the Creation of an Academic Space for Electrical Engineering in Britain." *History and Technology* 13: 73–111.

Guthrie, F. 1873. "On a Relation between Heat and Electricity." *Philosophical Magazine* 46: 257–266.

Hacking, Ian. 1992a. "The Self-Vindication of Laboratory Science." In *Science as Practice and Culture*, ed. A. Pickering. University of Chicago Press.

Hacking, Ian. 1992b. "'Style' for Historians and Philosophers." *Studies in History and Philosophy of Science* 23: 1–20.

Hall, H. C. 1902. "Wireless Telegraphy: To the Editor." *The Electrician* 50 (November 21): 198–199.

Hammond, J. H. Jr. and E. S. Purington. 1957. "A History of Some Foundations of Modern Radio-Electronic Technology." *Proceedings of the IRE* 45: 1191–1208.

Hancock, H. E. 1974. *Wireless at Sea*. Arno.

Heaviside, Oliver. 1892. *Electrical Papers*. 2 volumes. Macmillan.

Hijiya, James. 1993. *Lee De Forest and the Fatherhood of Radio*. A Lehigh University Press.

Hill, J. Arthur, ed. 1932. *Letters from Sir Oliver Lodge*. Cassell.

Hong, Sungook. 1994a. Forging the Scientist-Engineer: A Professional Career of John Ambrose Fleming. Ph.D. dissertation, Seoul National University.

Hong, Sungook. 1994b. "From Effect to Artifact: The Case of the Cymometer." *Journal of the Korean History of Science Society* 16: 233–249.

Hong, Sungook. 1994c. "Marconi and the Maxwellians: The Origins of Wireless Telegraphy Revisited." *Technology and Culture* 35: 717–749.

Hong, Sungook. 1995a. "Efficiency and Authority in the 'Open versus Closed' Transformer Controversy." *Annals of Science* 52: 49–76.

Hong, Sungook. 1995b. "Forging Scientific Electrical Engineering: John Ambrose Fleming and the Ferranti Effect." *Isis* 86: 30–51.

Hong, Sungook. 1996a. "From Effect to Artifact (II): The Case of the Thermionic Valve." *Physis* 33: 85–124.

Hong, Sungook. 1996b. "Styles and Credits in Early Radio Engineering: Fleming and Marconi on the First Transatlantic Wireless Telegraphy." *Annals of Science* 53: 431–465.

Hong, Sungook. 1996c. "Syntony and Credibility: John Ambrose Fleming, Guglielmo Marconi and the Maskelyne Affair." In *Scientific Credibility and Technical Standards in 19th and Early 20th Century Germany and Britain*, ed. J. Buchwald. Kluwer.

Hong, Sungook. 1998. "Unfaithful Offspring? Technologies and Their Trajectories." *Perspectives on Science* 6: 259–287.

Hong, Sungook. 1999. "Historiographical Layers in the Relationship between Science and Technology." *History and Technology* 15: 289–311.

Hounshell, David E. 1975. "Elisha Gray and the Telephone: On the Disadvantage of Being an Expert." *Technology and Culture* 16: 133–161.

Hounshell, David E. 1976. "Bell and Gray: Contrasts in Style, Politics, and Etiquette." *Proceedings of the IEEE* 64: 1305–1314.

Houston, E. J. 1884. "Notes on Phenomena in Incandescent Lamps." *Transactions of the American Institute of Electrical Engineers* 1: 1–8.

Howe, G. W. O. 1944. "An Interesting Patent Decision: Marconi's W. T. Co. of America versus the United States." *Wireless Engineer* 21: 253–255.

Howe, G. W. O. 1955. "The Genesis of the Thermionic Valve." In *Thermionic Valves 1904–1954: The First Fifty Years*. Institution of Electrical Engineers.

Hughes, Thomas P. 1966. "Introduction." In Samuel Smiles, *Selections from Lives of the Engineers*. MIT Press.

Hughes, Thomas P. 1983. *Networks of Power: Electrification in Western Society, 1880–1930*. Johns Hopkins University Press.

Hunt, Bruce J. 1983. "'Practice vs. Theory': The British Electrical Debate, 1888–1891." *Isis* 74: 341–355.

Hunt, Bruce. 1991. *The Maxwellians*. Cornell University Press.

Isted, G. A. 1991a. "Guglielmo Marconi and the History of Radio—Part I." *General Electric Company Review* 7: 45–56.

Isted, G. A. 1991b. "Guglielmo Marconi and the History of Radio—Part II." *General Electric Company Review* 7: 110–122.

Jackson, H. B. 1902. "On Some Phenomena affecting the Transmission of Electric Waves over the Surface of the Sea and Earth." *Proceedings of the Royal Society* 70: 254–272.

Jackson, Joseph Gray. 1970. "Patent Interference Proceedings and Priority of Invention." *Technology and Culture* 11: 598–600.

Jacot, B. L., and D. M. B. Collier. 1935. *Marconi, Master of Space: An Authorized Biography of the Marchese Marconi*. London.

Jenkins, Reese V. 1984. "Elements of Style: Continuities in Edison's Thinking." *Annals of the New York Academy of Science* 424: 149–162.

Johnson, J. B. 1960. "Contribution of Thomas A. Edison to Thermionics." *American Journal of Physics* 28: 763–773.

Jolly, W. P. 1972. *Marconi*. Constable.

Jolly, W. P. 1974. *Sir Oliver Lodge*. Constable.

Jordan, D. W. 1982. "The Adoption of Self-Induction by Telephony, 1886–1889." *Annals of Science* 39: 433–61.

Josephson, M. 1959. *Edison: A Biography*. McGraw-Hill.

Kelvin [William Thomson]. 1904a. "On Electric Insulation in Vacuum." *Philosophical Magazine* 8 (October): 534–38.

Kelvin [William Thomson]. 1904b. "Electric Insulation in a Vacuum." *The Electrician* 53 (October 14): 1031.

Kurylo, F., and C. Süsskind. 1981. *Ferdinand Braun: A Life of the Nobel Prizewinner and Inventor of the Cathode-Ray Oscilloscope*. MIT Press.

Lagergren, S. 1898. "Ueber die Dämpfung elektrischer Resonatoren." *Annalen der Physik* 64: 290–314.

Lamb, Horace. 1883. "On Electrical Motion in a Spherical Conductor." *Philosophical Transactions of the Royal Society* 174 (part II): 519–549.

Larmor, Joseph, ed. 1902. *The Scientific Writings of the Late George Francis FitzGerald*. Longman.

Laudan, Rachel. 1984. "Notice toward a Philosophy of Science/Technology Interaction." In *The Nature of Technological Knowledge*, ed. R. Laudan. Reidel.

Law, John. 1991. "Theory and Narrative in the History of Technology: Response." *Technology and Culture* 32: 377–384.

Leslie, Stuart W., and Bruce Hevly. 1985. "Steeple Building at Stanford: Electrical Engineering, Physics, and Microwave Research." *Proceedings of the Institute of Electrical and Electronic Engineers* 73: 1169–1180.

Lessing, Lawrence. 1956. *Man of High Fidelity: Edwin Howard Armstrong*. Lippincott.

Lewis, Tom. 1991. *Empire of the Air: The Men Who Made Radio*. HarperCollins.

Lodge, Oliver. 1889. *Modern Views of Electricity*. Macmillan.

Lodge, Oliver. 1890a. "Electric Radiation from Conducting Spheres, an Electric Eye, and a Suggestion regarding Vision." *Nature* 41: 462–463.

Lodge, Oliver. 1890b. "On Lightning-Guards for Telegraphic Purposes and on the Protection of Cables From Lightning." *Journal of the Institution of Electrical Engineers* 19: 346–379.

Lodge, Oliver. 1891a. "Some Experiments with Leyden Jars" (abstract). *Nature* 43: 238–239.

Lodge, Oliver. 1891b. "Presidential Address in Section A." *Report of the British Association for the Advancement of Science* (1891): 550–551.

Lodge, Oliver. 1892. *Lightning Conductors and Lightning Guards*. Longman.

Lodge, Oliver. 1894a. "Work of Hertz." *Proceedings of the Royal Institution* 14: 321–349.

Lodge, Oliver. 1894b. "The Work of Hertz." *Nature* 50: 133–139.

Lodge, Oliver. 1894c. *The Work of Hertz and Some of His Successors*. Van Nostrand.

Lodge, Oliver. 1897. "The History of the Coherer Principle." *The Electrician* 40: 87–91.

[Lodge, Oliver.] 1898a. "Dr. Lodge on Wireless Telegraphy." *Electrical Review* 42 (January 28): 103–104.

Lodge, Oliver. 1898b. "Improvements in Magnetic Space Telegraphy." *Journal of the Institution of Electrical Engineers* 27: 799–851.

Lodge, Oliver. 1900. *Signalling through Space without Wires* (third edition of Lodge 1894c). Electrician Publishing Co.

Lodge, Oliver. 1903. "Means for Electrifying the Atmosphere on a Large Scale." *The Electrician* 52 (November 20): 173–174.

Lodge, Oliver. 1905. "A Pertinacious Current; or, the Storage of High-tension Electricity by means of Valves." *Proceedings of the Royal Institution* 18: 79–86.

Lodge, Oliver. 1921–22. "Alexander Muirhead" (obituary). *Proceedings of the Royal Society* 100 (A): viii–ix.

Lodge, Oliver. 1923. "The Origin or Basis of Wireless Communication." *Nature* 111: 328–332.

Lodge, Oliver. 1925. *Talks about Wireless.* London.

Lodge, Oliver. 1926. "Reminiscences of the last British Association Meeting in Oxford, 1894." *Discovery* 7 (August): 263–266.

Lodge, Oliver. 1931. *Advancing Science.* London.

Lodge, Oliver. 1932. *Past Years: An Autobiography.* Scribner.

Lubell, Simon. 1942. "Magnificent Failure." *Saturday Evening Post,* January 17, 24, and 31.

MacGregor-Morris, J. T. 1954. *The Inventor of the Valve: A Biography of Sir Ambrose Fleming.* Television Society.

Marconi, Degna. 1982. *My Father Marconi.* Balmuir.

Marconi, Guglielmo. 1899. "Wireless Telegraphy." *Journal of the Institution of Electrical Engineers* 28: 273–291.

Marconi, Guglielmo. 1901. "Syntonic Wireless Telegraphy." *Journal of the Society of Arts* 49: 506–515.

Marconi, Guglielmo. 1902a. "Address." *The Electrician* 48 (February 21): 712–713.

Marconi, Guglielmo. 1902b. "Note on a Magnetic Detector of Electric Waves, which can be employed as a Receiver for Space Telegraphy." *Proceedings of the Royal Society* 70: 341–344.

Marconi, Guglielmo. 1902c. "A Note on the Effect of Daylight upon the Propagation of Electrodynamic Impulses over Long Distance." *Proceedings of the Royal Society* 70: 344–347.

Marconi, Guglielmo. 1902d. "The Progress of Electric Space Telegraphy." *Proceedings of the Royal Institution* 17: 195–210.

Marconi, Guglielmo. 1902e. "The Inventor of Wireless Telegraphy: A Reply." *Saturday Review* 93: 556–557.

Marconi, Guglielmo. 1903a. "Address." *Transactions of the American Institute of Electrical Engineers* 19: 98–101.

Marconi, Guglielmo. 1903b. "Address." *The Electrician* 50 (April 3): 1001–1002.

Marconi, Guglielmo. 1905. "Recent Advances in Wireless Telegraphy." *Proceedings of the Royal Institution* 18: 31–45.

Marconi, Guglielmo. 1908. "Transatlantic Wireless Telegraphy." *Proceedings of the Royal Institution* 19: 107–130.

Marconi, Guglielmo. 1909. "Wireless Telegraphic Communication." In *Nobel Lecture Physics* (Elsevier, 1967).

Maskelyne, Nevil. 1902. "A Supplement to Lieut. Solari's Report on The Radio-Telegraphic Expedition of H.I.M.S. 'Carlo Alberto.'" *The Electrician* 50 (November 7): 105–109.

Maskelyne, Nevil. 1903a. "Electrical Syntony and Wireless Telegraphy." *The Electrician* 51 (June 19): 357–360.

Maskelyne, Nevil. 1903b. "Wireless Telegraphy at the Royal Institution." *The Electrician* 51 (July 10): 503.

Maskelyne, Nevil. 1903c. "To the Editor." *The Electrician* 51 (July 17): 549.

Maskelyne, Nevil. 1903d. "Wireless Telegraphy at the Royal Institution." *The Electrician* 51 (July 24): 592.

Mason, Joan. 1991. "Hertha Ayrton and the Admission of Women to the Royal Society of London." *Notes and Records of the Royal Society of London* 45: 201–220.

Maxwell, James Clerk, J. D. Everett, and A. Schuster. 1876. "Report of the Committee for Testing Experimentally Ohm's Law." *Report of the British Association for the Advancement of Science* (1876): 36–63.

McCormack, Alfred. 1934. "The Regenerative Circuit Litigation." *Air Law Review* 5: 282–295.

McGrath, P. T. 1902. "Marconi and His Transatlantic Signal." *The Century Magazine* 63: 769–782.

Miessner, Benjamin Franklin. 1964. *On the Early History of Radio Guidance*. San Francisco Press.

Minchin, G. M. 1891. "Detection of Electro-Magnetic Disturbance at Great Distance." *The Electrician* 28 (November 27): 85.

Moffett, Cleveland. 1899. "Marconi's Wireless Telegraphy." *McClure's Magazine* 13: 99–112.

Moulton, Hugh Fletcher. 1922. *The Life of Lord Moulton*. Nisbet.

Muirhead, M. E. 1926. *Alexander Muirhead*. Oxford.

Nahin, Paul. 1988. *Oliver Heaviside: Sage in Solitude*. IEEE Press.

Nodon, A. 1904. "Electrolytic Rectifier: An Experimental Research" (Read at the St. Louis International Congress). *The Electrician* 53 (October 14): 1037–1039.

O'Dell, T. H. 1983. "Marconi's Magnetic Detector: Twentieth Century Technique Despite Nineteenth Century Normal Science?" *Physis* 25: 525–548.

Perry, John. 1910. *Spinning Tops: The "Operatives Lecture" of the British Association Meeting At Leeds, 6th September, 1890*. Society for Promoting Christian Knowledge.

Phillips, Vivian J. 1980. *Early Radio Wave Detectors*. Peter Peregrinus.

Phillips, Vivian. J. 1993. "The 'Italian Navy Coherer' Affair: a Turn-of-the-Century Scandal." *Institution of Electrical Engineers Proceedings* 140 (May): 173–185.

Pickering, Andy. 1993. "The Mangle of Practice: Agency and Emergence in the Sociology of Science." *American Journal of Sociology* 99: 559–589.

Pitt, Joseph C. 1988. "'Styles' and Technology." *Technology in Society* 10: 447–456.

Planck, Max. 1897. "Notiz zur Theorie der Dämpfung electrischer Schwingungen." *Annalen der Physik* 63: 419–421.

Pocock, Rowland F. 1963. "Pioneers of Radiotelegraphy." *Proceedings of the IEEE* 51: 959.

Pocock, Rowland F. 1988. *The Early British Radio Industry*. Manchester University Press.

Poincaré, H., and F. K. Vreeland. 1904. *Maxwell's Theory and Wireless Telegraphy*. Rob.

Post, Robert C. 1976. "Stray Sparks from the Induction Coil: The Volta Prize and the Page Patent." *Proceedings of the IEEE* 64: 1279–1286.

Poulsen, V. 1905. "System for Producing Continuous Electric Oscillations." In *Transactions of the International Electrical Congress, St. Louis*.

Poulsen, V. 1906. "A Method of Producing Undamped Electric Oscillations and Its Employment in Wireless Telegraphy." *The Electrician* 58 (November 16): 166–168.

Preece, William H. 1884. Discussion of Houston's "Notes on Phenomena in Incandescent Lamps." *Transactions of the American Institute of Electrical Engineers* 1: 1–8.

Preece, William H. 1885. "On a Peculiar Behaviour of Glow-Lamps when Raised to High Incandescence." *Proceedings of the Royal Society* 38: 219–230.

Preece, William H. 1893. "On the Transmission of Electric Signals Through Space," *Electrical World* 22 (September 2): 179–180.

Preece, William H. 1896a. "Electrical Disturbances in Submarine Cables." *The Electrician* 37: 689–691.

Preece, William H. 1896b. "On Disturbance in Submarine Cables." *Report of the British Association for the Advancement of Science* (1896): 732. (Title only).

Preece, William H. 1896c. "On the Transmission of Electric Signals through Space." *Electrical World* 22 (September 2): 179–180.

Preece, William H. 1897. "Signalling through Space without Wires." *The Electrician* 39 (1897): 216–218. Later published in *Proceedings of the Royal Institution* 15 (1896–97): 467–476.

Preece, William H. 1898. "Aetheric Telegraphy." *Journal of the Institution of Electrical Engineers* 27: 869–886.

Prescott, George B. 1888. *Electricity and the Electric Telegraph*. Appleton.

Price, Derek de S. 1984. "The Science/Technology Relationship, the Craft of Experimental Science, and Policy for the Improvement of High Technology Innovation." *Research Policy* 13: 3–20.

Pyatt, E. 1983. *The National Physical Laboratory: A History*. Hilger.

Ramsay, John F. 1958. "Microwave Antenna and Waveguide Technique before 1900." *Proceedings of the Institute of Radio Engineers* 46: 405–415.

Ratcliffe, J. A. 1974a. "Marconi: Reactions to his Transatlantic Radio Experiment." *Electronics and Power*, May 2: 322

Ratcliffe, J. A. 1974b. "Scientists' Reactions to Marconi's Transatlantic Radio Experiment." *Institution of Electrical Engineers Proceedings* 121, September: 1033–1038.

Reich, Leonard S. 1977. "Research, Patents, and the Struggle to Control Radio: A Study of Big Business and the Uses of Industrial Research." *Business History Review* 51: 208–235.

Reingold, N. 1959–60. "U.S. Patent Office Records as Sources for the History of Invention and Technological Property." *Technology and Culture* 1: 156–167.

Richardson, O. W. 1901. "On the Negative Radiation from Hot Platinum." *Proceedings of the Cambridge Philosophical Society* 11: 286–295.

Richardson, O. W. 1903. "The Electrical Conductivity Imparted to a Vacuum by Hot Conductors." *Philosophical Transactions* 201: 497–549.

Robinson, E. 1972. "James Watt and the Law of Patents." *Technology and Culture* 13: 115–139.

Rosenberg, Nathan. 1992. "Scientific Instrumentation and University Research." *Research Policy* 21: 381–390.

Rosling, P. 1906. Rectification of Alternating Currents." *The Electrician* 58: 677–679.

Rothmund, V., and A. Lessing. 1904. "Versuche mit dem elektrolytischen Wellendetektor." *Wiedemann's Annalen* 15 (September): 193–212.

Rowlands, Peter, and J. Patrick Wilson, eds. 1994. *Oliver Lodge and the Invention of Radio*. P.D. Publications.

Rowlands, Peter. 1990. *Oliver Lodge and the Liverpool Physical Society*. Liverpool: Liverpool University Press.

Rutherford, E. 1897. "A Magnetic Detector of Electrical Waves and some of Its Applications." *Philosophical Transactions* 189 (ser. A): 1–24.

Rutherford, E. 1902. "Marconi's Magnetic Receiver: To the Editor." *The Electrician* 49 (July 25): 502.

Schuster, Arthur. 1874. "On Unilateral Conductivity." *Philosophical Magazine* 48: 251–258.

Schuster, Arthur. 1890. "The Discharge of Electricity through Gases." *Proceedings of the Royal Society* 47: 526–561.

Seifer, Marc J. 1985. "Nikola Tesla: The Lost Wizard." In Tesla '84: Proceedings of the Tesla Centennial Symposium.

Selby, G. W. 1898. "Electric Telegraphy without Line Wires by Means of Hertz Waves." *The Electrician* 40 (January 14): 397–398.

Shapin, Steven, and Simon Schaffer. 1985. *Leviathan and the Air-Pump: Hobbes, Boyle and the Experimental Life*. Princeton University Press.

Sharp, Clayton H. 1921. *The Edison Effect and Its Modern Applications*. Privately printed in New York.

Shiers, G. 1969. "The First Electron Tube." *Scientific American* 220 (March): 104–112.

Slaby, A. 1898. "The New Telegraphy: Recent Experiments in Telegraphy with Sparks." *Century Magazine* 55 (April): 867–874.

Smiles, Samuel. 1904. *Lives of the Engineers.* Murray.

Smith, W. 1888. "Telegraphy without Wires: To the Editor." *The Electrician* 21 (November 2): 832–833.

Solari, L. 1902. "The Radio-Telegraphic Expedition of H.I.M.S. 'Carlo Alberto.'" *The Electrician* 50 (October 24): 22–26.

Spottiswoode, William. 1880. "Electricity in Transitu." *Proceedings of the Royal Institution* 9: 427–436.

Star, S. L., and J. R. Griesemer. 1989. "Institutional Ecology, 'Translations,' and Boundary Objects: Amateurs and Professionals in Berkeley's Museum of Vertebrate Zoology, 1907–39." *Social Studies of Science* 19: 387–420.

Stranges, A. N. 1986. Review of *Syntony and Spark* by Hugh G. J. Aitken. *American Historical Review* 91: 1166–1167.

Süsskind, Charles. 1962. "Popov and the Beginning of Radiotelegraphy." *Proceedings of the Institute of Radio Engineers* 50: 2036–2047.

Süsskind, Charles. 1968a. "The Early History of Electronics I. Electromagnetic before Hertz." *IEEE Spectrum* 5 (August): 90–98.

Süsskind, Charles. 1968b. "The Early History of Electronics II." *IEEE Spectrum* 5 (December): 57–60.

Süsskind, Charles. 1969a. "The Early History of Electronics III. Prehistory of Radiotelegraphy." *IEEE Spectrum* 6 (April): 69–74.

Süsskind, Charles. 1969b. "The Early History of Electronics: IV. First Radiotelegraphy Experiments." *IEEE Spectrum* 6 (August): 66–70.

Süsskind, Charles. 1970a. "The Early History of Electronics V." *IEEE Spectrum* 7 (April): 78–83.

Süsskind, Charles. 1970b. "The Early History of Electronics VI." *IEEE Spectrum* 7 (September): 76–84.

Süsskind, Charles. 1974. "Guglielmo Marconi (1874–1937)." *Endeavour* 33: 67–72.

Tattersall, James J. and Shawnee L. McMurran. 1995. "Hertha Ayrton: A Persistent Experimenter." *Journal of Women's History* 7: 86–112.

Taylor, J. E. 1903. "Characteristics of Earth Current Disturbances and their Origin." *Proceedings of the Royal Society* 71: 225–227.

Tesla, N. 1894. "On Light and Other Frequency Phenomena" (lecture delivered before the National Electric Light Association, St. Louis, 1893), in *The Inventions, Researches and Writings of Nikola Tesla*, ed. T. Martin. McCall & Emmet.

Thackeray, D. 1984. "Edison's Electrical Indicator." *Electronics and Wireless World* 90 (February): 30–31.

Thompson, Jane S., and Helen G. Thompson. 1920. *Silvanus Phillips Thompson: His Life and Letters.* Dutton.

Thompson, Silvanus P. 1887–08. "The Development of the Mercurial Air-Pump." *Journal of the Society of Arts* 36: 20–49.

Thompson, Silvanus P. 1896. "On the Property of a Body Having a Negative Electric Resistance." *The Electrician* 37 (July 3): 316–318.

Thompson, Silvanus P. 1897. *Light, Visible and Invisible.* Macmillan.

Thompson, Silvanus P. 1898. "Telegraphy Across Space." *Journal of the Society of Arts* 40: 453–460.

Thompson, Silvanus P. 1902a. "The Inventor of Wireless Telegraphy." *Saturday Review* 93 (April 5): 424–425.

Thompson, Silvanus P. 1902b. "Wireless Telegraphy: A Rejoinder." *Saturday Review* 93: 598–599.

Thompson, Silvanus P. 1910. *The Life of William Thomson.* Macmillan.

Thompson, Silvanus P. 1911. *Notes on Sir Oliver Lodge's Patent for Wireless Telegraphy.* Privately printed in London.

Thomson, Elihu. 1899. "The Field of Experimental Research." *The Electrician* 43: 778–780.

Thomson, J. J. 1883–04. "On Electrical Oscillations and the Effects Produced by the Motion of an Electrified Sphere." *Proceedings of the London Mathematical Society* 15: 197–219.

Thomson, J. J. 1891. "On the Discharge of Electricity through Exhausted Tubes without Electrodes." *Philosophical Magazine* 32: 321, 445.

Thomson, J. J. 1893. *Recent Researches.* Cambridge University Press.

Thomson, J. J. 1902. "On Some Consequences of the Emission of Negatively Electrified Corpuscles by Hot Bodies." *Philosophical Magazine* 4: 253–262.

Threlfall, Richard. 1890. "The Present State of Electrical Knowledge." *Report of the Australasian Association for the Advancement of Science* 2: 27–54.

Trouton, F. T. 1892. "Radiation of Electric Energy." *The Electrician* 28 (January 22): 301–303.

Trowbridge, John. 1897. *What Is Electricity?* Kegan Paul, Trench, Trübner.

Tucker, D. G. 1981–02. "Sir William Preece (1834–1913)." *Transactions of the Newcomen Society* 53: 119–138.

Tyne, G. F. J. 1977. *Saga of the Vacuum Tube.* H. W. Sams.

Van Helden, Albert, and Thomas L. Hankins, eds. 1994. *Instruments. Osiris,* vol. 9.

Vendik, Orest G. 1995. "Popov, Marconi and Radio." *Nature* 374 (April 20): 672.

Vyvyan, R. N. 1933. *Wireless over Thirty Years.* Routledge.

White, Lynn, Jr. 1962. "The Act of Invention: Causes, Contexts, Continuities, and Consequences." *Technology and Culture* 3: 486–500.

White, W. C. 1943. "Electronics . . . Its Start from the 'Edison Effect' Sixty Years Ago." *General Electric Review* 46 (October): 537–541.

Wilson, Adrian. 1993. "Foundations of an Integrated Historiography." In *Rethinking Social History: English Society 1570–1920 and Its Interpretation,* ed. A. Wilson. Manchester University Press.

Wilson, David. 1983. *Rutherford: Simple Genius*. Hodder and Stoughton.

Wise, Norton M. 1988. "Mediating Machines." *Science in Context* 2: 77–113.

Woodbury, David O. 1931. *Communication*.

Woodbury, David O. 1944. *Beloved Scientist: Elihu Thomson*. Boston Museum of Science.

Woodruff, A. E. 1966. "William Crookes and the Radiometer." *Isis* 57: 188–198.

Yavets, Ido. 1993. "Oliver Heaviside and the Significance of the British Electrical Debate." *Annals of Science* 50: 135–173.

Index

Printed in the United States
by Baker & Taylor Publisher Services